W0230257

Green China

Contemporary China faces serious environmental problems which have been widely documented in the western media, usually accompanied by doom-laden assessments and predictions for the future.

This is the first book to locate China's environmental problems in the context of her rapid industrialisation, urbanisation and transition to a market-oriented economy. Drawing on a wide range of Chinese and western sources, the book offers in-depth analysis of the complete range of environmental problems facing China today, from the historical, political, economic and cultural root causes, through the successful and unsuccessful efforts which have been made to find solutions, to possible future scenarios and strategies.

Rejecting the blanket pessimism of other studies of the Chinese environment, this book offers a constructive perspective on a subject frequently dealt with in overwhelmingly negative terms.

Geoffrey Murray has worked in Pacific Asia for 40 years. He is a Research Associate of the Centre for Pacific Rim Studies, Liverpool John Moores University.

Ian G. Cook is Professor of Human Geography and Head of Geography and the Centre for Pacific Rim Studies at Liverpool John Moores University.

Green China

Seeking ecological alternatives

Geoffrey Murray and Ian G. Cook

RoutledgeCurzon
Taylor & Francis Group

LONDON AND NEW YORK

First published 2002 by RoutledgeCurzon
This edition published 2013 by Routledge
2 Park Square, Milton Park, Abingdon, Oxon, OX14 4RN

Simultaneously published in the USA and Canada
by Routledge
711 Third Ave, New York NY 10017

Routledge is an imprint of the Taylor & Francis Group

Transferred to Digital Printing 2005

© 2002 Geoffrey Murray and Ian G. Cook
First issued in paperback 2013
Typeset in Goudy by Steven Gardiner Ltd, Cambridge

British Library Cataloguing in Publication Data
A catalogue record for this book is available from the British Library

Library of Congress Cataloging in Publication Data
A catalog record for this book has been requested

ISBN 978-0-700-71703-3 (hbk)
ISBN 978-0-415-86163-2 (pbk)

Contents

Tables

Boxes

Preface

Attending the China Environmental Forum in November 1997, various speakers queued up to warn that China's industrial boom was causing acid rain, poisoned oceans and global warming – all problems that could create regional tensions if not addressed quickly.

Former Australian Prime Minister Paul Keating, for example, said the fact China burnt coal to supply 80 per cent of its energy needs meant it was pumping some 10 million tonnes of sulphur into the air every year; the sulphur and other pollutants spouting from hundreds of thousands of ageing mainland factories were at the root of about half of the acid rain falling on East Asia. 'China's carbon dioxide emissions threaten to become a major cause of global warming. Left unchecked, environmental problems are likely to become a source of tension, and these problems, unless suppressed, spill over into political and strategic problems', he warned.

At the same forum, Taiwan voiced fears that pollution dumped into the ocean by the heavily industrialised Yangtze River could decimate fish stocks, while Japan was particularly worried about smog from China riding the wind to return to earth as acid rain. Mitsutake Okano, a senior adviser to Mitsubishi Corp., declared: 'Our concerns on acid rain are coming from the continent, including China'. While the issue had not yet led to bilateral bickering, the Japanese government had held talks with its Chinese counterpart on solving the problem and was encouraging Japanese companies to bring clean technology to China, he said.

In the same year, Zhao Bin, writing in the *New Left Review*, warned that China's 'furious industrialisation' was 'fast propelling it toward the dubious distinction of being the world's Number One polluter' (Zhao, 1997, p. 17).

Meanwhile, among some 350 scientists attending the launch of the biggest study of ozone loss over Europe and the Arctic at a research centre in Kiruna, northern Sweden in January 2000, British scientist Joe Farman, the man who discovered the ozone hole over the Antarctic in 1985, also identified China as a threat to global well being. The failure of China and India to phase out dangerous chemicals threatened the progress made in healing the earth's protective ozone layer, he warned. Since his initial discovery, the volume of ozone-killing substances released into the atmosphere had been cut, but more needed to be done, and 'the problem is now with the developing countries, especially China, India and perhaps Malaysia and Indonesia as well' (Reuters, 24 January 2000). Coincidentally, three days later China announced its first list of controlled ozone-depleting substances as part of a pledge to follow international norms aimed at protecting the critical atmospheric layer (*China Daily*, 28 January 2000).

The cross-border mobility of the mainland's pollution was also brought home through a two-year study by scientists in the Institute of Environment and Sustainable Development of Hong Kong's University of Science and Technology, who found pollutants clearly identifiable as coming from the Gobi Desert 2,000 kilometres away. Checks with counterparts in Taiwan revealed the same phenomenon causing 'yellow rain' to fall over the island. Institute director Dr Fang Ming said the particles, probably carried by the jet stream in the upper atmosphere, were high in calcium, sulphur and iron, characteristic of the soil in northern China (Cheung Chi-Fai, 2000).

The 'export' of mainland pollution even became the subject of political jousting between Hong Kong and its mainland neighbour Guangzhou, over the issue of whether the former's appalling air quality was home grown or came from the latter's cement factories. Guangzhou might be the most seriously polluted city in the Pearl River Delta, conceded Mayor Lin Shusen, but its air quality was still much better than that of the Hong Kong Special Administrative Region. 'How can Guangzhou's fresher air affect Hong Kong's worsening sky?', he asked in an address to the city's Eleventh People's Congress.

Demonstrating his knowledge of chemistry, the politician said exhaust fumes from cars, buses and trucks contained nitride and oxide, while cement factories predominantly discharged gas containing carbon monoxide, carbon dioxide and dust. The vehicles on its roads were responsible for at least 80 per cent of the pollutants in Hong Kong's air. Situated at the mouth of Pearl River, and facing the South China Sea, it enjoyed a much better geographical position than inland Guangzhou, but its many skyscrapers prevented the exhaust gases from dispersing. 'And that is why Guangzhou residents can breathe fresher air than their friends and relatives in Hong Kong', the Mayor concluded triumphantly (*China Daily*, 4 April 2000).

The relative merits of Guangzhou's and Hong Kong's approach to clean air will be discussed later in Chapters 4 and 9 respectively, and can be set aside for the moment. The purpose of these brief vignettes is to point up China's increasingly dominant influence in determining the success or failure of the regional (at the very least), and possibly even the global effort to deal with the environmental devastation caused by rapid industrialisation, growing consumerism and soaring populations in the twentieth century. They will be central to the arguments we will be putting forward in this book.

In the opening chapter, we briefly examine the seriousness of the problem as seen by various authoritative sources, both within China and elsewhere.

We look at the main contributory factors, such as a heavy emphasis on industrialisation to promote China's economic growth, its burgeoning population and the worrying trend towards Western-style consumerism of a growing proportion of that population. We place this within both a political and a historical context, especially the attitude under Communism that nature was there to be tamed and altered at political whim. We then consider what achievements have been made in recent years towards reversing the trend of environmental degradation, while at the same time pointing out the heavy cost in economic terms of any serious clean-up and the difficulty of achieving this while still maintaining high economic growth.

We have set out to provide a fairly exhaustive study of the problems facing China today, as well as considering both existing remedies and possible scenarios for the future. As we make clear at the very start of Chapter 1, these are issues of global concern, for China's dominant geographical position on the Asian landmass create influences that are widely felt beyond its borders – acid rain devastating forests across the seas in Japan, for example. Not only the 1.3 billion people of the People's Republic of China, but an even greater number in neighbouring countries from India to Korea, desperately need to see viable solutions created that will not only halt further eco-environmental degradation on the mainland, but create significant improvements in air, earth and water quality.

It is not our intention to castigate China. One must have some sympathy with the government in Beijing as it tries to deal with problems spread over such a vast territory of differing climatic and geological conditions, as well as trying to educate its huge population to be more ecologically aware. One should also recall the incredible backwardness of the country in 1949, when the late Chairman Mao Zedong and his followers took control of the country after some thirty years of war. China's Communist rulers have done much to create a powerful industrial base that has given the vast majority of Chinese a better standard of living than the masses have ever known. But the price for this in terms of environmental damage has been extremely high, and cannot be allowed to continue.

We have, therefore, set out to take a candid look at China's problems without being overly condemnatory. But we do consider it absolutely vital that there is a healthy debate of all the issues so that workable solutions can be found. Throughout the book we have sought to present data that identifies both the seriousness of a problem and the probable causes, while at the same time acknowledging, where justified, China's efforts to achieve an improvement.

We have approached the task with a certain amount of scepticism, due to the tendency of the Chinese authorities to exaggerate the good news aspects of any situation, while downplaying the bad. There has, for example, been a tendency for the government to claim a problem has been solved, only to have it return in an even worse form almost immediately (for instance, the dust storm problem in Beijing discussed in Chapter 1).

But, as shown by Mr Xie's comments in Chapter 1 below, there is now a tendency for more openness and admissions that things are indeed bad. Until quite recently, to cite one example, one would only discover that there had been a major coalmine disaster with heavy loss of life when, months later, the state media carried a eulogistic story on the wonderful advances that had been achieved in industrial safety thanks to the forward-looking attitude of the Communist Party since the aforementioned disaster. Now, bad news is likely to be treated in the same way is it would in the West – with big front-page headlines. Until quite recently, one could not have found statements such as the following: 'Like many other countries, China has many [environment] problems. In some badly polluted areas, poisoned vegetables have killed some children; healthy pregnant women give birth to deformed babies; and the number of cancer cases caused by pollution has risen' (www.china.org.cn 2001). Such candour can only be good for the environment, for only when every citizen of China is convinced that things are bad and that 'something must be done about it' is there likely to be strong, measurable improvement.

Nevertheless, this remains a touchy subject. Despite increasing candour about its problems, China continues to resent outside criticism if it sees this as 'interfering in China's internal affairs'. Thus, while pledging Chinese determination to 'take the initiative in international environment diplomacy', SEPA minister Xie Zhenhua could, in a major speech in June 2000, also stress that 'we oppose those using environmental problems as a pretext for intervening in other countries' sovereignty and interior affairs'.

China had always held that developing and developed countries should both shoulder the responsibility for protecting the environment, but considered both historically and realistically, developed countries were the most responsible for the current environmental problems, while developing countries were the victims. Therefore, developed countries should be the first to take action, while in the meantime helping developing countries participate in campaigns for environmental protection. Developed countries should not impose pressure on developing countries or ask them to take on unreasonable duties at international environmental conferences or treaties parleys, Mr Xie declared.

With this in mind, we have sought to draw from a wide range of Chinese and Western sources, personal observation and interviews in China a picture that is fair to China while avoiding falling into over-optimistic, propagandist excess. Even so, some in the academic community invited to review the first draft of this work criticised us for alleged excessive reliance on 'propaganda' (in other words, government statements carried in the still-dominant official media). This, to us, seemed a throwback to the old Cold War days, and failed to recognise, as already stated, that not only the official media but also even the government itself are no longer wedded to the idea that China is already a socialist utopia. In addition, we would suggest that the Chinese media today is no more guilty of spreading 'propaganda' than are the major newspapers in

Britain when they publish reports by their journalists who have just had a background briefing from a government 'spin doctor'. So we make no excuse for quoting from the state-run media, whether it is reporting good or bad news.

It is our hope that the finished work will be of interest to both the specialist and the general reader who wants to understand the complexities of the current battle on the mainland between economic advancement and eco-environmental destruction. After identifying the key issues in the opening chapter, we provide two chapters of historical context, dealing with the Chinese people's relationship to the environment throughout history, and the drive to dominate nature that typified the reign of Chairman Mao in the mid-twentieth century (e.g. the Great Leap Forward) and the disasters that resulted. The two following chapters are then devoted to the chief factors contributing to the current level of environmental deterioration, notably large-scale industrialisation and the growing urban demographic consumerist pressures resulting from the increasing wealth industrialisation has generated.

In Chapters 6 and 7, we return to the theme of Maoist-style challenges to nature, this time in the current era, by considering, first, the Sanxia (Three Gorges) Dam project, and then the plans to divert water from the mighty Yangtze River to its ailing northerly companion, the Yellow River. Chapter 8 looks at other current environmental problems, while Chapter 9 concentrates on three areas on the mainland periphery – Hong Kong, Taiwan and Tibet.

In Chapter 10, we look at what is being done now to tackle the various problems previously discussed, at levels ranging from the local up to the international, before rounding out the book with a consideration of various future scenarios ranging from the relatively optimistic to that of blackest gloom. Our feeling is that China will end up somewhere between these two stark pillars. We would like to think that this chapter and the work as a whole will help create better understanding of China's problems and help identify what needs to be done.

This collaboration comes about from the shared long-time interest of the two authors in China – adding up to thirty years of research experience – cemented through work together at the Centre for Pacific Rim Studies, Liverpool John Moores University, and collaboration on the book *China's Third Revolution: Tensions in the Transition to Post-Communism*. As with that work, we have divided the labour equally between us. Each author produced an initial draft for some of the chapters, based partly on personal interest and knowledge, which were then made available for further editing and additions by other person. This, we believe has worked well, for we come from different backgrounds – Ian Cook from academia, Geoffrey Murray originally from international journalism – and have been able to contribute a rich vein of professional personal experience to the final product.

Ian Cook would like to acknowledge with thanks the kindness of Liverpool John Moores University in granting him sabbatical leave in the first semester 1999–2000 in order to enable him to work on the book. The help of his colleagues in Geography and Politics also deserves much appreciation. On his part, Geoffrey Murray would like to thank various friends in Beijing in helping him to understand the existing situation and stay balanced.

Liverpool/Beijing June 2002

1 China's environmental crisis: an overview

'Let trees sprout on the mountains; stop growing grain on hilly terrain; and keep livestock in their pens'; thus, the new environmental creed for the twenty-first century, according to Chinese Premier Zhu Rongji. This edict, a classically structured piece of propaganda consisting of three balanced lines of four characters each, reportedly emerged from the premier's inspection tours of northern and north-eastern provinces in the spring of 2000. In those regions, lack of green spaces exacerbated the fierce sand storms blowing through Beijing and contributing to a serious drought in the capital, of which Mr Zhu, and every other resident, was only too well aware.

Reporting on the results of his trip to the country's top administrative body, the State Council, the premier was remarkably frank in laying the blame firmly at the ambitious but short-sighted agricultural practices of Chairman Mao Zedong (see Chapter 3), in instructing farmers to terrace hilly areas and plant rice or wheat. The new edict from Zhu specifically forbade the very things that Mao championed, and was reportedly to be accompanied by handouts of free grain to enable farmers to plant acres of trees rather than food crops, and free coal to stop them from burning trees and shrubs for fuel.

'Every year, we spend a lot of money deploying aircraft to seed the clouds for rain – and to sow seeds for grass and trees in hilly regions,' Mr Zhu reportedly told a meeting of agronomists in Beijing. 'Yet, farmers continue to ask their kids to dig out barely grown sods for fuel or forage' (*People's Daily*, 21 April 2000).

There was, of course, no guarantee that unscrupulous farmers would not take the subsidies and continue to grow crops or scrape the hills for fuel. Nevertheless, it seems to us, that it will surely be cash, rather than any catchy edict, which will be the key, given that one of great failings of government policy on the environment so far has been the failure to achieve widespread grass-roots involvement in any clean-up work. For no matter how much money one puts into propaganda campaigns, no programme will work without economic incentives – making it worthwhile for people to become involved.

As yet this does not seem to be the case. As the Chinese American academic Dr Thomas Shen observes:

> Millions of Chinese talk about pollution, but [most] have little idea about pollution. Many people see environmental problems, but still many others do not. Even those who see environmental pollution problems may not know and understand what happened, why it happened and how it happened.
>
> (Shen, T.S., 2001)

But this must change if there is to be significant reverse in the widespread environmental degradation now facing China. Marking World Environment Day China 2000, Xie Zhenhua, minister of the State Environmental Protection Administration (SEPA), admitted that while the trend to environmental degradation had been stopped or even reversed in parts of the country, prospects for environmental protection were grim.

As a legacy of the country's headlong dash for economic growth (discussed in Chapter 3 and 4), about 28,000 enterprises still failed to meet the state environmental standards. Another 7,280 had failed to take action to reduce their discharge of toxic wastes despite repeated warnings to do so, the minister reported. Industrial pollution can occur due to mismanagement caused by a variety of reasons, such as 'carelessness, indifference or ignorance, as well as lack of measurement and monitoring methods to provide baseline or background data. In China, these problems are legion' (Shen, op. cit.).

Addressing the Innovation Strategy Forum sponsored by the Chinese Academy of Sciences on 26 April 2000, Xie Zhenhua presented a stark picture of existing problems, along with a few small victories his ministry had achieved:

- **Preliminary control of pollutant discharge**. By the end of 1999, total discharge of 12 main industrial pollutants basically decreased from 1995 levels of 1995; despite steady economic development being maintained, discharge volumes of some main pollutants were down by 15–20 per cent from 1995. The total discharge of sulphur dioxide was down from 24.6 million tons in 1995 to 18.57 million tons in 1999, and the emission of soot dropped from 17.5 million tons in 1995 to 11.59 million tons in 1999.
- **Change in pollutant mix**. Pollution caused by vehicle exhaust gases was even more severe in urban areas, and most prominent in large cities such as Beijing, Shanghai, Wuhan and Guangzhou. About 50 per cent of nitrogen oxide pollutants in Beijing are caused by vehicle exhaust emissions, replacing coal fires as the main pollutant source.
- **Mixed results in dealing with water pollution**. The proportion of industrial pollution was down, but that of domestic pollution up; in many places, domestic pollution equalled the level of industrial pollution. Farming activities account for 20 per cent of the total water pollution sources, the heavier pollution in lakes consisting of

phosphorus and nitrogen. The level of pollution by toxic and harmful substances that do not degrade is worsening. Studies of drinking water sources in some large cities found more than 20 different carcinogenic substances.

• **Deteriorating ecology.** Land deterioration is increasing; since the founding of the People's Republic of China (PRC) in 1949, the area of soil erosion nationwide has reached 3.67 million square km, accounting for 38 per cent of the nation's total land area. In addition, this area is expanding at an annual rate of 10,000 sq. km. Deserts have been advancing and the frequency of sandstorms is increasing; vegetation has been seriously destroyed, and desert and alkaline grassland accounts for one-third of the total land area.

• **Water ecosystem disturbed.** Disruptions in the water ecosystem have led to frequent flood disasters, river run-offs, dried-up lakes (see Box 1.1), destruction of wetlands and the decline of underground water.

• **Bio-diversity destroyed.** Of the higher plants, 15–20 per cent are in peril of extinction, greater than the 10 per cent world average level.

Having presented this overview of mainly bad news, Mr Xie went on to identify the short-term tasks for environmental protection as follows:

1 In maintaining steady economic development, total pollutant discharge is to be reduced by another 10 per cent by the end of 2000.
2 Efforts to be made toward the conservation of water resources, protecting the ecological environment at the Yangtze (especially the Three Gorges Reservoir area and its upper reaches) and the Yellow (especially the Xiaolangdi River and its upper reaches) river valleys.
3 In the next five years, the number of key cities for environmental protection is to be raised to 100.
4 The content of sulphur dioxide in control areas is to be cut by another 20 per cent by the end of 2000.
5 Measures are to be taken to protect the environment in ecology-fragile areas, resource development areas and nature reserves.

A year later, the minister went even further in outlining the ecological and environmental protection programme for the ensuing decade. He pledged the government would spend around one per cent of national GDP each year of the Tenth Five-Year Plan (2001–5). Table 1.1 Outlines the government's goals for this period.

In ten years, the minister added, the government aimed to raise forest and brushwood coverage in desert areas to 11.83 per cent to control the spread of sands (www.china.org.cn 2001, op. cit.).

The SEPA minister's outline provides an excellent starting point for anyone wishing to understand the problem of environmental and ecological degradation in the People's Republic of China as it enters the twenty-first

---

Box 1.1 Disappearing lakes

An average of 20 natural lakes dry out in China every year. According to the National Forestry Bureau, large-scale enclosure of tideland for cultivation and the disturbance of surface water flow have caused a rapid shrinkage in lake water. In eastern and central China, nearly 1,000 natural lakes have disappeared in the past 50 years due to land reclamation; Hubei province had 1,052 lakes in the 1950s, but only 83 are left now. In western China, water is scarce and evaporates quickly. With the large-scale interception and diversion of surface water flow, some major lakes are gradually deteriorating into salty or dry salt lakes. Aydingkol Lake and many others are already lifeless deserts. Manas Lake in the western part of the Junggar Basin used to have a surface area of 577.8 sq. km. In recent years, however, the continuous diversion of water for irrigation has stopped the flow of water from rivers into the lake, turning it and the surrounding salt marsh and grassy marshland into dry saline and bleak desert. (Xinhua 12 January 2001)

---

century. Any visitor to the country today does not have to go far to experience this first-hand. The encroaching deserts, for example, are evident in the appalling dust storms that sweep over Beijing every spring (a phenomenon described in detail later in this chapter). The blanket of smog that hovers over the majority of the mainland cities for most of the year provides ample evidence of both the heightened industrial activity in one of the world's economic powerhouses and the advance of the motor vehicle to the core of Chinese life.

Less obvious to the casual visitor are the massive economic losses China suffers through environmental damage, and the health problems that are emerging due to bad air and bad water. These will form an important thread running right through every chapter of this book.

Progress at high cost

Since the economic reforms of 1978, China has experienced dramatic industrialisation and rising energy use against a backdrop of population growth and unprecedented urbanisation. Its astounding industrial growth over the past two decades has created a country poised to become a major economic power in the twenty-first century (for a fuller discussion see Murray (1998) and Cook and Murray (2001)). Per capita, it is still one of the world's poorest countries, yet the future looks promising as incomes rise, poverty rates fall, and life expectancy reaches Western levels. But, along with these gains, it is grappling with environmental problems that could prevent it from sustaining high levels of economic growth in the coming decades.

Table 1.1 Major government goals for the Tenth Five-Year Plan

Item	Specific goals
Overall expenditure on environmental protection	700 billion yuan, 1.3% of expected GDP (0.93% in Ninth Five-Year Plan).
Industrial pollution	Discharges cut by at least 10% on average nationwide
Urban pollution	Over 100 cities to meet state standards for water/air quality and noise levels. Clean fuel technology for industry and vehicles further promoted.
CO emissions and acid rain	About 72% of cities in the south-west to meet state second-level standard for CO; acid rain greatly reduced.
River/lake water	Wastewater treatment plants on Huaihe, Haihe and Luohe Rivers, Taihu, Chaohu and Dianchi Lakes; south–north diversion project.
Coastal waters	Treating water entering Bohai Sea; all coastal waters to meet state standard.
General	Raising public awareness through a national telephone complaint centre.

This is a crucial issue for the entire global community, given China's geographic, economic and demographic presence. Encompassing a geographically vast area, with a number of distinct ecological zones, it extends from the massive and sparsely populated Gobi Desert in the north and the mountains of the south-western Himalayas to the densely inhabited valleys of the eastern coast. As the world's most populous country (almost 1.3 billion counted in the 2001 census), its economic growth is the fastest and most sustained of any major country in the world, at an average of 10 per cent annually over more than a decade (World Bank, 1997a), although some of the most prosperous areas along the east coast and in southern China have achieved more than double this figure (Cook and Murray, op. cit., Chapter 1).

Industry is the largest productive sector, accounting for 48 per cent of gross domestic product (GDP) and employing 15 per cent of the country's total labour force (International Labour Organisation, 1996, p. 205). In the 1990s, the output of China's 10 million industrial enterprises has increased by 18 per cent annually (World Bank, op. cit., p. 130), and this is undoubtedly the key factor in lifting many millions of people out of poverty. It also underlies a huge and growing demand for energy, with coal still dominant (probably 75 per cent for the foreseeable future by our estimations) despite the growing use of oil and natural gas.

Along with industrialisation (see Chapter 4) has come rapid urbanisation, especially in what is known as the southern coastal crescent that runs from Guangzhou to Shanghai (see Chapter 5). The proportion of the population living in cities has grown about 50 per cent since 1980. Some 370 million people now live in cities, and this number is expected to grow to 440 million by the turn of the century (United Nations, 1997). One World Bank model predicts that, by the year 2020, 42 per cent of China's population, more than 600 million people, will live in urban areas overwhelmingly concentrated in the eastern and southern coastal provinces (Li *et al.*, 1995, p. 17).

Drawing from a variety of Chinese government sources, including annual environmental reports, plus assessments by international organisations such as the World Bank, the key problems facing China today can be summarised as follows:

- Over half the population (nearly 700 million people) lacks access to clean water, and consumes drinking water contaminated with animal and human waste exceeding maximum permissible levels. Some 400 Chinese cities are regarded as short of water, while more than 100, including Beijing, are 'acutely short'. Excessive over-extraction of ground water, meanwhile, is causing cities such as Shanghai, Tianjin and Xian to subside.
- The State Oceanographic Bureau (SOB) says rising sea levels around China threaten to destroy dykes built in the past to now outdated engineering demands. Excessive extraction of underground water in a number of seaside cities has considerably lowered local ground water levels and caused land to cave in, facilitating seawater invasion (SOB, 2001a).
- Each year, factories discharge some 36 billion tons of untreated industrial wastewater and raw sewage into rivers, lakes and coastal waters. Overall, only 5 per cent of household waste and about 17 per cent of industrial waste receive any treatment before entering local irrigation ditches, ponds, lakes, and streams. Nearly half of the country's rivers and over 90 per cent of its urban water resources have been polluted to some extent. Serious pollution has been documented in the country's seven major watersheds: Huai, Hai, Liao, Songhua, Chang (Yangtze), Zhu (Pearl) and Huang (Yellow); each year, increasingly longer stretches of the Yellow River dry up for longer periods (see Chapter 7).
- There are severe deforestation problems (contributing heavily to the devastating floods in the summer of 1998, for example). In a country where only about 10 per cent of the land is cultivatable, a third of cropland has been lost since 1960 from over-farming (soil erosion, desertification), energy projects and to industrial and housing projects. Land lost to deserts increased by an average 2,460 sq. km yearly in the 1980s and 1990s.
- Air pollution in some cities averages more than ten times the standard proposed by the World Health Organisation, threatening public health and welfare on a large scale. China's six largest cities – Beijing, Shenyang, Chongqing, Shanghai, Xian, and Guangzhou – rank among the most polluted in the world. According to the former National Environmental Protection Agency, polluted air and respiratory disease had become the

leading cause of death in urban areas and even some rural areas (Liu Yinglang, 1996), although to be fair, heavy smoking is another important contributory factor to the latter. Mercury and lead poisoning, meanwhile, was increasingly detected in children throughout the 1990s (WRI, 1999b).

• If present trends continue in the emission of greenhouse gases, particulates and heavy metals, by the early years of the twenty-first century China will be the world's largest producer of acid rain and largest emitter of greenhouse gases.

Amid all the gloom, China's coastal waters offer a comparative bright spot. According to the State Oceanographic Bureau, in 2000 the marine environment stopped deteriorating for the first time in years, offering hope for the future. It said the general pollution situation of the sea was 'roughly the same as' instead of the familiar 'worse than' that of the previous year. The area of polluted coastal waters, one of the most important indications of marine pollution as a whole, was still around 260,000 sq. km, with 29,000 sq. km classified as 'seriously' polluted, meaning that marine life and harbour operations are impossible. Discharges from the land, which contain rich oils, organic substances and heavy metals, are still the leading sources of pollution, especially off the country's advanced land areas, like Shanghai and Tianjin as well as the provinces of Zhejiang, Liaoning, Jiangsu and Guangdong (SOB, 2001a).

While cataloguing a generally negative picture, however, one is bound to observe that in 1949 the new People's Republic of China faced a massive burden of nutritional deficiency and infectious and parasitic diseases. More than half the population died as a result of infectious and other non-degenerative diseases before reaching middle age. In the past half century, the average life span has doubled to 70, while the infant mortality rate has dropped from 200 per 1,000 to 31 per 1,000. Infectious diseases, while still a serious problem in some parts of the country, claim the lives of a mere 0.0004 per cent of the population each year (Ministry of Public Health, 1996, p. 3). These are remarkable achievements, and due credit should be given to the aggressive campaign to improve primary health care and tackle infectious diseases.

Yet, in the coming decades, there is concern that a deteriorating environment could well undermine the gains that rising incomes would otherwise bring. The government itself has also identified a deteriorating environment as a major factor influencing the morbidity and mortality rates (NEPA, op. cit.), while a 1994 opinion survey of experts with science or engineering degrees ranked risk from pollution ahead of natural disasters (Zhang Jianguang, 1994, p. 165).

There is concern that the chronic environmental problems of the urban areas are spreading to the countryside and ecological destruction is intensifying (NEPA, 1997). Throughout the 1990s, environmental problems

Table 1.2 Losses resulting from pollution 1992

Environmental factor	Value of economic loss (billion yuan)	Per cent of total loss
Water pollution	35.6	36.1
Human health	19.28	
Industry	13.78	
Crop yields	1.38	
Livestock	0.7	
Fisheries	0.46	
Air Pollution	57.89	58.7
Human health	20.16	
Agriculture	7.2	
Home upkeep	13.44	
Clothing	1.06	
Vehicles	1.07	
Buildings	0.96	
Acid Rain	14	
Solid waste	5.12	5.2
Total	98.61	100

Source: World Resources Institute (1999), *World Resources 1996–7.*

seriously affected the country's overall social and economic development, the economic costs associated with ecological destruction and environmental pollution being estimated as high as 14 per cent of the gross national product (Chen Qian, 13 September 1997, p. 1). The World Bank, meanwhile, has put the annual cost of air and water pollution at around US$54 billion, or nearly 8 per cent of the Chinese GNP (World Bank, 1997b, p. 23). Another source noted that, while the government had invested 2,500 billion yuan in controlling man-made and natural disasters in the 50-odd year history of the PRC, annual losses caused by these same events had reached 283 billion yuan (www.china.org.cn, 2001, op. cit.).

The figures in Table 1.2 are somewhat dated, but they provide an interesting breakdown of how the various forms of pollution impact on a broad range of human activity.

Consumerist contribution

These costs are likely to become matters of even greater concern in the coming years if China maintains its headlong pursuit of capitalist-style consumerism (see Chapter 5). As Zhao Bin argues, 'perhaps nowhere is the impact of the transition to capitalism having a more devastating effect than upon China's environment'. (Zhao, op. cit., p. 13). Citing the case of Taiwan, where a panel of experts in 1989 warned that the island's rapid development – often hailed in the West as evidence of capitalism's superiority over communism – risked turning it into a 'poisonous garbage dump', he asserts:

'Given the huge scale of China's transition to capitalism, this has ominous implications, and not only for the Chinese' (ibid., p. 14). We discuss Taiwan's problems more fully in Chapter 9.

Zhao goes on to say that, on a per capita basis, the billion residents of the developed world in the 1990s consumed at least

> three times as much water, 10 times as much energy 13 times as much iron and steel, 14 times as much paper, 18 times as much chemicals and 19 times as much aluminium as someone in a developing country like China. Industrial countries account for nearly two-thirds of the global emissions of carbon dioxide from the combustion of fossil fuels [and] their factories generate most of the hazardous chemical wastes. Their air conditioners, aerosol sprays and factories release almost 90 per cent of the chlorofluorocarbons that destroy the ozone layer.
>
> (Ibid., p.15)

Imagine then, what will be the affect of adding *another* billion consumers, which is what is gradually going to happen if China maintains its present move towards general affluence. The question must be asked: can China, or indeed the global environment, sustain the load this huge augmentation of mass consumption will impose.

According to the US-based World Resources Institute (WRI), now engaged in a long-term co-operative program with the Chinese Government on ways to balance economic growth with environmental protection, over the past 100 years the United States and other highly industrialised countries, have contributed 65 per cent of the concentrations of carbon dioxide above the pre-industrial baseline. Even though China's population is three-and-a-half times larger than that of the United States, the latter has contributed four-and-a-half times more carbon dioxide to the atmosphere this century:

> On a per capita basis, Americans emit 7.5 times more carbon dioxide than the average Chinese. And, most significantly, a recent report on the Chinese electric power sector by Battelle Memorial Institute and two Chinese research centres estimates that since the late 1970s, China's economy grew twice as fast as its energy consumption. China would be emitting twice as much carbon dioxide into the atmosphere if not for an unprecedented 'de-coupling' of carbon emissions and economic growth.
>
> (WRI, 1999a)

According to data compiled by the United States Energy Department, throughout the 1990s, China's carbon dioxide emissions from the burning of fossil fuels increased at an annual rate of 0.9 per cent, compared with 1.3 per cent for the United States. 'There is a good basis to argue that China has done more to combat climate change over the past decade than the United States', declared the Natural Resources Defence Council, an American environmental group. Meanwhile Californian researchers found that China had achieved 'stunning' and 'dramatic' success in reducing carbon dioxide emissions, saying they had *shrunk* 17 per cent since the mid 1990s, while the national GDP had grown 36 per cent (*New York Times* 17 June 2001). Asked about this, Zhou Dadi, director of the Chinese Energy Research Institute, claimed that the country's per capita energy use was 'just one-tenth of that in the US and one-seventh of that in Europe' (Xinhua 18 June 2001).

Nevertheless, China *is* catching up. And the per capita argument tends to obscure the real threat it poses. One of the authors (Murray) can recall attending an international environmental conference in Beijing in 1991, where Chinese speakers sought first, to paint Western attempts to get China to curb its pollution levels as a desire to sabotage the country's development as an economic competitor, and second, to argue that, *on a per capita basis*, pollution levels were still very low compared to the West. Given that China has more than four times the population of the United States (and likely to have close to five times before population growth finally flattens out) this seemed somewhat irrelevant. In addition, China is achieving in one decade the sort of economic growth that it took the Western developed world a century to achieve. In *absolute terms*, China is already the third biggest contributor to global pollution and is on course to become world leader – contributing, for example, 40 per cent of the world's carbon dioxide emissions in 2050.

Who should pay?

The Chinese leadership tends to see the solution of environmental problems in terms of the developed world taking much of the responsibility for action, on the grounds that they have been the main beneficiaries of the economic advancement that has had pollution as an unfortunate side effect. There is still a significant segment of the government and the Communist party that wants to continue putting economic growth ahead of environmental concerns (partly, it would seem, in the belief that the party can only continue to command a semblance of loyalty among the general citizenry if it continues to 'deliver the goods' in economic terms), and who firmly oppose any attempts to impose limitations on China's growth for the sake, say, of cleaner air in the developed countries.

This attitude was perfectly encapsulated in the following statement by Li Junfeng, a senior energy researcher for the State Planning Commission, attending the First Conference of the Parties to the Framework Convention on Climate Change, held in Berlin in April 1995. He told delegates bluntly:

> It is just as hot in Beijing as it is in Washington DC. You try to tell the people in Beijing that they can't buy a car or an air conditioner because of the global climate-change issue. If we reduce our emission of gasses it

means we must reduce our energy consumption. When people get rich, they want to buy an air conditioner or a car; that will increase energy consumption.

But, while one may have some sympathy with the idea of China and all other developing countries having the perfect right to catch up to the economic level long enjoyed by the developed world, unhappily the realities of global environmental deterioration no longer make this practical. At the same time, however, it does seem somewhat hypocritical for a country like the Unites States to criticise China, while continuing its long-term excessive consumption of the world's resources.

But politics aside, the fact is that, as the Greenpeace organisation has pointed out, China faces 'environmental meltdown' if it waits to get rich before tackling its numerous grave environmental ills. A 32-page report by Greenpeace China issued in 1999 observed that a country

> which uses seven times more energy to produce one dollar of gross domestic product than developed countries, must promote cleaner and more efficient energy use and not wait until it gets rich to clean up its skies and waters. If we have to pollute first and clean up later, that means we are paying twice.

Admittedly, solving these problems will not come cheap, and, while professing to be a 'developing country', one has to ask whether China could have done more from its own resources. For example, the government spent an estimated US$950 million on average each year between 1991 and 1995 on environmental clean-up and protection works. The Ninth Five-Year Plan, meanwhile, budgeted a total of US$6.75 billion, or US$1.35 billion a year, for projects to deal with air and water pollution control, solid waste disposal and noise abatement. In addition, the World Bank provided some US$500 million a year during the same period. But this is far from enough to meet the country's needs. With healthy trade surpluses and foreign exchange reserves now approaching US$200 billion, could not the government have done more to tackle the issue itself without waiting for international handouts?

And, yet, that seems to be a significant part of the Chinese strategy. Attending the 1997 China Environmental Forum referred to earlier (in the Preface), then Premier Li Peng called on overseas companies to help foot the bill for the huge mainland ecological clean-up. 'I hope those powerful international businesses will provide financial support to developing countries to help them deal with environmental problems,' he declared.

The official English-language media picked up the tune and began carrying many articles drumming home the message of the wonderful opportunities, and profits, to be made by investing in environment-related projects. For example, foreign investors in waterworks projects were guaranteed a certain percentage of the inevitable profits for 15 or 20 years.

Sustainability theory

But beyond all the arguments about who is to blame and who is to pay, the larger issue, as we will repeatedly show in ensuing chapters, is one of 'sustainability'. Can, for example, China maintain for an extended period its current determined drive for high economic growth year after year, in order to ensure a reasonable standard of prosperity for all its present citizens, without leaving a legacy of waste and loss to blight the lives of its future citizens? Are the grandiose schemes for re-ordering nature to deal with historic natural disasters such as floods merely creating the conditions for even worse disasters in future?

As this is at the core of the ongoing debate, we need to look at how sustainability is defined in an international context. First, we accept the idea that environmental policy cannot be limited to warding off imminent hazards and repairing the damage that has already occurred (an important element of the Chinese catch-up effort described throughout this book). What is needed is an environmental policy in which natural resources are protected and demands on them are made with care, and political and commercial decisions that will have an environmental impact in any shape or form are only made after exhaustive checks to ensure they will be in no way detrimental to the ecological balance now and in the foreseeable future.

One approach to this is the 'Precautionary Principle', first developed in Germany in the 1970s, and perhaps best defined by the group of scientists, government officials, lawyers, and environmental representatives that met at Wingspread in Racine, Wisconsin in 1998. The group came up with the Wingspread Statement:

> The release and use of toxic substances, resource exploitation, and physical alterations of the environment have had substantial unintended consequences on human health and the environment. Some of these concerns are high rates of learning deficiencies, asthma, cancer, birth defects and species extinctions; along with global climate change, stratospheric ozone depletion; and worldwide contamination with toxic substances and nuclear materials. We believe existing environmental regulations and other decisions, particularly those based on risk assessment, have failed to adequately protect human health and the environment, as well as the larger system of which humans are but a part. While we realize that human activities may involve hazards, people must proceed more carefully than has been the case in recent history.

The key elements of the Precautionary Principle are:

1   People have a duty to take anticipatory action to prevent harm.
2   The burden of proof of harmlessness of a new technology, process, activity, or chemical lies with the proponents.

3   Before using a new technology, process, or chemical, or starting a new activity, people have an obligation to examine 'a full range of alternatives' including that of doing nothing.
4   Decisions applying the precautionary principle must be 'open, informed, and democratic' and 'must include affected parties'.

Opponents of the Precautionary Principle stress that it is impossible to scientifically anticipate *all* the potential effects of a new technology or discovery. In particular, the pressures of the market economy, which rewards speed and market dominance, pose an important barrier. Nevertheless, the themes of the Precautionary Principle have been incorporated into many environmental plans and international treaties, including: the Rio Declaration (Principle 15, 1992), the Helsinki Convention (1992), and the Framework Convention on Climate Change (1992).

Another way of dealing with sustainability is offered by the 'Natural Step', a philosophy emerging from Scandinavia in the 1980s to provide a set of guidelines that address ecological, economic and social concerns. The four conditions it sets are:

1   In order for a society to be sustainable, nature's functions and diversity are not systematically subject to increasing concentrations of substances extracted from the earth's crust. In a sustainable society, human activities such as the burning of fossil fuels and the mining of metals and minerals will not occur at a rate that causes them to systematically increase in the ecosphere, leading eventually to global warming, contamination of surface and ground water, and metal toxicity In practical terms, this means comprehensive metal and mineral recycling programs, and a decrease in economic dependence on fossil fuels.
2   In order for a society to be sustainable, nature's functions and diversity are not systematically subject to increasing concentrations of substances produced by society. In a sustainable society, humans will avoid generating systematic increases in persistent substances such as DDT, PCBs, and Freon gas. Society needs to find ways to reduce economic dependence on persistent human-made substances.
3   In order for a society to be sustainable, nature's functions and diversity are not systematically impoverished by physical displacement, over-harvesting or other forms of ecosystem manipulation. In a sustainable society, humans will avoid taking more from the biosphere than can be replenished by natural systems. In addition, people will avoid system-atically encroaching upon nature by destroying the habitat of other species. Biodiversity, which includes the great variety of animals and plants found in nature, provides the foundation for ecosystem services necessary to sustain life on this planet. Society's health and prosperity depends on the enduring capacity of nature to renew itself and rebuild waste into resources.

4   In a sustainable society resources are used fairly and efficiently in order to meet basic human needs globally. Meeting the fourth system condition is a way to avoid violating the first three system conditions for sustainability. Considering the human enterprise as a whole, we need to be efficient with regard to resource use and waste generation in order to be sustainable.

A third approach to sustainability comes from the CERES Principles, enunciated by the Coalition for Environmentally Responsible Economies, a non-profit organisation formed in the United States and comprising major corporations and investor, environmental, religious, labour and social justice groups. Formerly known as the Valdez Principles, stemming from the environmental disaster caused by the 1989 Exxon Valdez oil spill in Alaska, the significance of the CERES Principles mainly lies in establishing a sustainability dialogue in the business community, enabling it to evaluate existing managerial styles and adopt sustainable management practices.

These principles point to a renewed sense of corporate responsibility to their employees, their communities and the environment. In addition, the ecological, economic and equality components of sustainability are no longer viewed as distinct but rather as complementary – the choice is not economic growth at the expense of the environment, but a vibrant economy, equitable resource distribution, and environmental protection. In addition, these principles outline guidelines that show companies how they can make a profit through socially just policies and safeguarding the environment. We would contend that this aspect has been lost sight of by a good many mainland local government officials, who place promotion of the local economy – and the incidental strengthening of their own power base – above all else

Whether we specifically refer to any of these principles of sustainability or not, they undoubtedly guide us in the ensuing chapters in drawing up a report card for China's protection of the environment.

Legacy of Han grandeur

It is only fair to say that, despite an obsession with grandiose schemes challenging nature in a disastrous way (see Chapter 3 on Politics in Command, Chapter 6 on the Three Gorges Dam and Chapter 7 on the proposed diversion of major rivers), not all the blame for ecological degradation can be laid at the door of the PRC's rulers.

As we will show in Chapter 2, over many centuries, the Han Chinese steadily migrated outward from their traditional home around the Yellow River to settle border areas – sometimes in search of a more prosperous or peaceful life, but often sent out by their rulers for strategic reasons. They brought along with them agricultural practices that were not always suitable for local conditions. In the grasslands of the north-west, for example,

Han farmers were constantly forced to move on because their ploughing exposed sandy layers that drifted away in the wind, and eventually led to desertification.

The French explorer Andre Migot, on a visit to the Sichuan capital of Chengdu in 1946, observed that 'the boundless, pullulating energy of the people tends to obliterate everything; historic monuments and ancient ruins have a shorter life than they have elsewhere' (Migot, 1955, p. 54). He was referring to the wholesale destruction of the country's architectural heritage, but the comment is equally apposite in relation to the environment as a whole.

A 1992 report by the Chinese Academy of Sciences admitted that there was a long history of irrational exploitation of natural resources and of environmental destruction:

> Due to abundant natural resources and a vast territory, the Chinese people engaged in frequent migrations, opening up new areas for settlement, creating a residue of damage from their primitive agricultural civilisation. Until 1949 when New China was born, the quality of China's ecological environment had declined to an historically unprecedented low level owing to plunder and wartime destruction and man's reckless exploitation and utilisation of natural resources. The excessive reclamation of land, overgrazing and deforestation over thousands of years caused serious degradation of forests, grasslands and cultivated land as well as severe soil erosion and desertification.
>
> (CAS, 1992, pp. 31–2)

The academy then felt obligated to insist that much had been done to rectify this situation after the PRC was founded in 1949. Much of this book, however, will inevitably question that claim. In Chapter 3, for example, we point out that 'Maoism', the all-inclusive ideology of the first three decades of the PRC was based partly on the materialist Marxist view that the earth could be controlled via human intervention.

Although the phrase 'politics in command' was developed to show command over the economy, we believe it was extended to command over the environment. Much of the environmental degradation described in opening paragraphs of this chapter can be laid mainly at the door of the Maoist insistence of economic development dominated by heavy (polluting) industry – an extension, it would seem, of the nineteenth-century British belief that 'where there's muck there's brass'.

The communisation of agriculture, with its emphasis on an industrial approach to crop growing, obsession with high output targets regardless of local soil or weather conditions, and the 'Four Pests Campaign', had a number of positive environmental features, especially with regards to water conservancy (see Chapter 3). At the same time, it contributed greatly to an ecological imbalance – further aggravated by massive tree felling to open up more land for development (subsequently blamed for devastating floods in the Yangtze Valley, for example, in the 1990s).

These shortcomings of official policy – whether they could have been anticipated or not – therefore, have to be taken into account when drawing up any balance sheet of 'socialist achievements'. And it should be stressed that, despite bad experiences, the post-Mao leadership has also shown an attachment to huge civil engineering schemes to overcome the perceived defects of nature. Thus, as already mentioned, there is the Three Gorges Dam, where even some Chinese environmentalists have spoken out about the dangers it may pose to the ecological balance to the area along the country's most crucial waterway. There is also another dream of Mao Zedong – a massive transfer of waters from the lush lands of southern China to drought-prone areas in the arid north by diverting water from the Yangtze to the Yellow River, famous for its disastrous floods throughout history but now in danger of drying up – that has now been taken off the shelf and dusted down for implementation. The arguments for and against this particular project are examined in Chapter 7.

But whatever the rights and wrongs of these nature-altering projects past and present, the fact is that the central government *has* responded to growing public concern by officially naming the environment as one of its top priorities and has committed itself to reversing the trend of environmental deterioration (Ninth Five-Year Plan [1996–2000] for Environmental Protection and Long-Term Targets to 2010).

In the 1990s, the government increased environmental spending, adopted market incentives, strengthened lawmaking and enforcement, and promoted national environmental education. We consider what has been done so far in Chapter 9. In addition, decisions made in the next decade or two about energy, transportation, and agricultural technologies will largely determine how successful China will be in achieving its goal of sustainable development. These are considered at various points in Chapters 3, 5 and 8, and in the final chapter.

One further encouraging development is that the fight for a cleaner environment is no longer exclusively a government preserve (to be stepped up or cut back as dictated by current political concerns). In 1998, for example, non-communist politicians severely attacked Beijing's record on environmental protection, forcing the government on the defensive.

After the 1998 Yangtze Valley floods, members of Chinese People's Political Consultative Conference (CPPCC) castigated National People's Congress President Li Peng for a 'lacklustre' performance on the environment during his ten-year premiership that had just ended, and urged his successor Zhu Rongji to do far more in the areas of irrigation, agriculture and flood prevention. The critics even went so far as to charge that the late Deng Xiaoping's reforms begun in 1978 were based on fast growth at the expense of the environment. Such criticism was unprecedented from what was generally regarded then as a powerless, rubber-stamp body, but was a warning to the government that the environmental issue was beginning to arouse passions (Lam, 1998).

Delegates to the Guangdong People's Congress, meanwhile, turned an otherwise staid meeting into a gruelling session for the province's environmental boss, blaming his department for Guangdong's rapidly deteriorating environment. Delegates from Foshan censured the director of the local environmental protection bureau, after they complained that 17 upstream electroplating factories on the Beijiang River had contaminated their city's water supplies. The director admitted his department had failed to assess the damage before allowing the factories to start up. 'Since they [the factories] have not had their environmental assessment reports completed [before operation], they should have been stopped,' he confessed. 'But some of my colleagues have approved them to start production. I should admit that we were wrong,' he said, adding that he would ask the factories temporarily to suspend operation. His reply did not appease the delegates. In a resulting vote, 23 delegates said they were dissatisfied and only five supported the bureau (Ma, 2000).

Achievements so far

Le us, however, acknowledge that much has been done around the country, and programmes have been put in place that will hopefully bring further improvements in the years ahead. Both aspects will be discussed in detail in later chapters, but at this stage we can summarise them as follows:

- **Air pollution**. The State Environmental Protection Administration (SEPA) predicted a 14 per cent cut in sulphur dioxide emissions by the end of 2000 and a 42 per cent reduction by 2010, based on programmes drawn up by prefectural and city governments in acid rain and sulphur dioxide control zones. The State Council demanded that each prefecture and city in the zones must draw up programmes to reach national standards on sulphur dioxide discharges by 2010, requiring the elimination of more than 10 million tons of sulphur dioxide emissions. Factories must either adopt clean energy sources and install pollution control facilities or close down. Meanwhile, major cities like Beijing have driven older polluting vehicles off the road and required new ones to have advanced emission control devices. The sale of leaded petrol has been banned. In addition, 53 small coal-fired power stations, and 6,400 coalmines have been closed down.
- **Water pollution**. An estimated 50,000 factories have been closed due to their waste discharges polluting the nation's chief waterways. In major cities like Beijing and Shanghai, a wholesale shift of industry from urban areas to the outer suburbs and beyond has been underway since 1998. Under the 'Three Simultaneous Requirements' (TSR), waste treatment facilities must be designed, constructed and operated coincidentally with

the new construction projects as a means of tackling industrial pollution from the start. Projects failing to conduct environmental-impact assessments and meet TSR requirements will not get approval to operate. Those found to have subsequently violated the requirements face heavy fines and closure.

- **Aridity**. The Ministry of Agriculture acknowledges a pressing need for China to develop agriculture that uses water efficiently in a bid to ensure security of food supplies and alleviate poverty in its vast arid areas. Ministry figures indicate half the country's arable land is afflicted by a water shortage, and 43 per cent of the population lives in arid regions. Pilot schemes are underway in the north of the country, the worst affected region, to restore desert and other poor quality land to useful purposes. Reforestation programmes on a vast scale are underway to provide 'shelter belts' to hold back the desert and alleviate the dust storms that have long plagued northern cities like Beijing. Logging operations in many parts of the country are now strictly controlled to curb soil erosion.

- **Energy Conservation**. The Energy Conservation Law that came into force on 1 January 1998, and subsequent regulations based on it, give the government stronger powers to prohibit certain new industrial projects that seriously waste energy and employ outmoded technologies. It covers energy from coal, crude oil, natural gas, electric power, coke, coal gas, thermal power, biomass power, and other energy sources. A determined drive has been underway since the mid-1990s to deal with the high sulphur content of much of the coal used throughout the country, including washing and other treatments, as well as building pithead power stations to keep the pollution away from urban areas. Alternative energy sources are being pursued, especially hydroelectric power. Pipelines now bring natural gas to Beijing and other major cities from as far away as neighbouring countries in Central Asia to further alleviate the need for using coal, and more were being built at the time of writing.

These four examples demonstrate that the main problems are being tackled *to some extent*. Whether these efforts are sufficient will be the subject of detailed analysis in subsequent chapters.

Need for caution

But one also needs to approach Chinese claims of successes with some caution. Under communism, there has been a tendency for the media – until recently, entirely state-run – to highlight the good news, even if the truth has to be stretched somewhat, and suppress the bad. Only in the latter half of the 1990s were there signs of a more independent-minded media industry willing to probe and publicise the nations numerous problems – including those related to the environment.

On occasions, the convolutions of the state media to put a positive spin on the news can almost be laughable. Elsewhere in this book we refer to the two-week-long blanket of smog that descended on Beijing within hours of the local media ecstatically reporting the permanent return of blue skies to the capital, thanks to the wise environmental policies of the Communist Party, the central and local governments.

Another perfect example is the dust storms that have long been an annual spring torment for Beijing's residents. At that time of year, bitterly cold winds sweeping south from Siberia and Mongolia tend to pick up sand from the deserts in northern China and deposit a considerable amount of it onto the city. In the mid-1990s, while living in Beijing, the author (Murray) recalls frequent reports in the state media of how the planting of a forest 'shelter belt' (dubbed a 'Green Great Wall') along with projects to reclaim parts of the desert for arable use meant the end of the sandstorm scourge. This quickly turned out to be nonsense, as he discovered on a visit to Beijing in April 2000, because the storms had actually become more severe.

Nevertheless, the state media took the change in its stride. The *China Daily*, which has certainly contributed much to the 'good news on the environment' phenomena, apparently saw nothing wrong in doing a complete about face – informing its readers at the height of the spring 2000 battering that 'dust storms have increased in China over the past five decades. Statistics note 5 sand storms annually in the 1950s, 8 in the 1960s, 13 in the 1970s, 14 in the 1980s and 23 in the 1990s' (*China Daily* 5 April 2000).

Between March and early April 2000, in fact, eight significant sand or dust storms battered the capital. The suffusing dust blocked out the sun, turned the sky slate grey and completely obscured the landscape. The ratio of airborne particles reached the standards of heavy air pollution. The Beijing Meteorological Station was even forced to add sandstorm forecasts to the original weather forecast. Winds as high as force nine were reported, which in one case led to three workmen being killed when scaffolding on a construction site was swept away. To ensure safety, many construction sites had to suspend work that was high above the ground. About 1,500 km of railways, 30,000 km of highway and 50,000 aqueducts were reported damaged to varying degrees, and direct economic losses from the storms were estimated at 54 billion yuan.

Similar sandstorms appeared in some regions in the eastern part of north-west China, the south-western part of north-east China and the northern part of north China. According to statistics of the State Meteorological Centre, between March and April Inner Mongolia, Hebei, west Liaoning and some regions in the northern part of north China all saw northerly winds of force 5 to 7, and sometimes even force 8 to 10, accompanied by sandstorms.

Professor Qiu Guoqing, who works at the centre, observed that compared with the same period of previous years, the latest sandstorms had a higher frequency, an earlier starting time, a broader range, and greater intensity. Ci Longjun, a professor with the Chinese Academy of Forestry, meanwhile, told

Table 1.3 Grassland degradation in source regions of Yangtze and Yellow Rivers (Measurement unit: 10,000 hectares)

Region	Total area	% of total grassland	Slight degradation	%	Moderate degradation	%	Heavy degradation	%
Darlag	51.04	45.69	10.60	20.76	25.51	49.98	14.91	29.22
Madoi	107.02	46.55	72.28	67.54	5.08	4.47	29.66	27.72
Magen	30.08	29.86	16.59	55.16	8.84	29.39	4.65	15.45
Zhidoi	43.27	20.17	33.08	76.45	4.19	11.34	5.28	2.14
Qumarleb	101.97	26.44	68.53	67.21	14.12	13.84	19.32	18.95

Source: Adapted from Wang and Cheng (2000), *The Environmentalist* 20, p. 223.

the *People's Daily* that northern China's dusty weather has been increasingly severe since 1997.

One sandstorm permeated as far south as Nanjing, capital of East China's Jiangsu Province, in April 1998, bringing suspended particles to the air that were eight times higher than normal. Ci said the increasing desertification in northern China had contributed a lot to the dusty weather, citing official statistics showing nearly 40 per cent of China's territory already desert, with another 2,460 sq. km added on average each year.

Although not directly related to Beijing's dust problems, this might be a good point to digress slightly to mention the work done by the Geology Department of Lanzhou University on the source regions of the Yangtze and Yellow Rivers. This provides us with some interesting figures on just how serious the desertification problem has become (Table 1.3 and 1.4) in these environmentally sensitive areas. Desert land in the source region of the Yellow River is mainly distributed in Madoi and Magen Counties, while 80 per cent of the desertified land in the Yangtze River source region is distributed in Zhidoi and Zadoi counties.

Yang Guimin, senior engineer with the China Central Meteorological Observatory, estimates that a million tons of sand is blown into Beijing annually from the area of Zhangjiakou, Hebei Province, about 160 km north-west of the capital. Guo Liang a researcher with the Beijing Municipal Meteorological Observatory, meanwhile, said the more than 5,000 construction sites in Beijing, its chronic scarcity of rain and dry air, and exposed soil with immature plants, added a further layer of dust to that sweeping in from the desert. Ding Yihui, a research fellow at the State Meteorological Centre, however, insisted this made only a minor contribution; it was an abnormal spring, with premature thawing of frozen soil, that was the main factor.

To reach Beijing, the winds from the north have to cross the Alxa Desert. Once a land of milk and honey, Alxa is now a sand storage area dominated by two huge dried lakes. Luo Bin, an engineer with the Desertification Prevention and Control Centre under the State Forestry Bureau, explained that in the 1950s the water volume of the Heihe River flowing from Qinghai and Gansu provinces into the Alxa region was 0.9 to 1.1 billion cu. m,

Table 1.4 Desertification in source regions of Yangtze and Yellow Rivers
(Measurement unit: hectare)

Region	Desertified land area	Moving dune	%	Semi-fixed and fixed dune	%
Zhidoi	176470	43844	24.83	111532	63.10
Madoi	72690	47630	65.28	25330	34.72
Magen	57700	57700	100	0	0
Darlag	1250	0	0	1250	100
Zadoi	17751	7219	40.67	10532	59.33
Total	322131	152393	47.31	148644	46.14

Source: Adapted from Wang and Cheng (2000), *The Environmentalist* 20, p. 224.

whereas by the late 1990s, the figure was 0.2 to 0.3 billion. With widespread elimination of vegetation, over 400,000 hectares of land became desert; of 353,000 ha of natural vegetation, over 23,000 ha had withered.

The ecological environment surrounding the capital showed similar deterioration, Luo claimed. According to official surveys, desert areas now constitute 64 per cent of the total land in the north-west Beijing periphery. The main reservoirs providing potable water for Beijing are Guanting, Miyun and Panjiakou. Their upstream areas, namely the Sangganhe River, the Chaohe River, the Baihe River and the Luanhe River, are all within the boundary of Hebei Province, where one tenth of the land is jeopardized by desertification. Over the past few decades, the sand belt along the Bashang area and the Sangganhe River has formed a patch of desert only 70 km away from Beijing, from where the sand is blown into the city. In Zhangjiakou, Hebei Province, a 14,000-ha desert now contributes nearly a million tons of sand to Beijing each year.

The appalling conditions now prevailing in the country's political and administrative centre have provoked a remarkable degree of candour on the seriousness of the problem, at odds with the traditional communist desire to look on the bright side of life. 'To some extent, the climate is beyond the control of human beings. But had there been vegetation and protective forests, wind and sand would have lost their foothold, and the precipitation would have been larger. The sandy winds caused by the destruction of vegetation is retribution for human wantonness', admitted Luo Bin. Meanwhile, Wang Zhibao, Director of the State Forestry Bureau, conceded:

China's ecological environment is facing a severe trend. The general trend of deterioration is not yet fully under control. At present the desert land continues to expand, totalling an area of 2.62 million ha and greatly surmounting that of the arable land in China. A 4,500 km-long, 600 km-wide sand belt is formed from the Tarim Basin in the west to the Songhuajiang–Nenjiang Plain in the east. The desert areas in the Xinjiang

Uygur Autonomous Region and Inner Mongolia account for 47 per cent and 60 per cent of their respective total land area. Deserts threaten many provinces and autonomous regions, such as Gansu, Qinghai, Ningxia, Shaanxi, Shanxi, Hebei, Liaoning, Jilin and Heilongjiang. China is now among the countries most severely jeopardized by desertification.

Experts at the Desertification Prevention and Control Centre attributed the acceleration in desertification to a combination of excessive land reclamation, excessive grazing, excessive lumbering, excessive gathering of herbs and desert plants, and excessive use of water.

Starting in the 1950s, the central government made what it called 'relentless efforts' to prevent and control the spread of sand. By 1998 it claimed to have reclaimed 70,000 sq. km of land for some form of arable use, with 30 million ha of grass and trees being planted in various shelter belt projects. This sparked the self-congratulatory stories mentioned above. Two years later, the mood was different, with admissions that the 'the general situation of land desertification is still worsening; the pattern of sand forcing people to retreat remains the same as before, and the plight is still severe'.

Yang Youlin, an expert at the State Forestry Bureau, admitted that the construction of the shelterbelt was slower than destruction, and the main reason lay in administration. Local governments had not properly carried out their duties, leading to excessive reclamation, lumbering and water exploitation. Another problem was the lack of funds. Most of the desert-threatened places lie in the west, where subsistence is still the priority, rather than environmental protection.

In the new mood of sober assessment, Yang's bureau has worked out a scheme for sand prevention and control stretching over the next half century. The scheme is divided into three stages. By 2010, they hope to control the comparatively dangerous but easily controlled sand sources. By 2030, they want to have throttled the pace of desertification. And by 2050, they aim to have raised national forest coverage to 26 per cent, thus bringing the entire sand source under effective control.

One wishes them well, for it will not be an easy problem to solve. Returning to Beijing in Spring 2001, the author (Murray) found the dust storms were, if possible, even more severe than the previous year and lasted beyond the seasonal norm, a pattern repeated again in 2002.

In discussing the cause of Beijing's growing dust storms in such detail at this point, we have been motivated by a number of considerations. First, we feel it illustrates the complexity both of the causes and the proposed countermeasures, and shows that dealing with China's numerous environmental problems will be no easy task. Despite the long-held belief that 'man can conquer nature', as expressed in numerous engineering projects ever since Mao Zedong came to power (see Chapters 3, 6 and 7 for prime examples), nature has a nasty habit of hitting back and making things worse.

Second, we wanted to provide a basis for some of the scepticism that will be expressed in subsequent chapters about any claims of success in dealing with China's environmental problems. However, the remarkable media about-turn noted above does offer us some encouragement that the days of excessively highlighting the positive and glossing over the negative, much beloved by Mao and his immediate successors, are gone so that China and its neighbours can really get to grips with the environmental issues we shall examine in detail in what follows.

Third, the difficulties in dealing with the advancing desert also help highlight some of the frustrations faced by the central government, which can issue the orders but cannot always make them stick at the local level. As we will consider in due course, localism and regionalism are two of the most difficult hurdles to be overcome in ensuring ecological balance.

2   Ancient legacies

It is probably no exaggeration to state that of all the peoples of the earth, the Chinese have the greatest sense of history, reflecting the greater continuity of Chinese civilisation compared to other heartlands of human occupancy and endeavour. Mao himself made great use of the ancient classics, for instance in his concept of guerrilla warfare, and for exemplars to inspire and direct the people. The sense of China's 'historical shame', the humiliation inflicted by Western dominance in the nineteenth and first half of the twentieth century, still plays an important part in Chinese perceptions and international relations. Notwithstanding the progress of recent decades, the legacies of the past still play a major role in contemporary affairs.

Some legacies, however, may be ignored or almost deliberately forgotten, for ideological and other reasons. This can be to the detriment of contemporary policies, as we shall show in later chapters. Here we shall examine some of the key legacies that have affected China's interaction with the environment, in terms of water, land and philosophy. The next sections look at the first by focusing on 'China's Sorrow', the Yellow River, 'Han Expansion – Desertification and Deforestation', and 'Ecological Balance' respectively.

China's Sorrow – the Yellow River and the river dragons

Throughout China's long history, water has played a major part in defining what the country would become. 'A hydraulic civilisation', the term developed by Karl Wittfogel in the early twentieth century, became for many years a convenient concept to summarise Chinese society and Chinese life. There was a perennial struggle with river water, often successful – but water cannot easily be channelled and tamed. Despite the successes, every so often, swollen with the summer rains, the rivers would burst through the dykes and flood huge areas, killing hundreds of thousands of people. 'The river dragon has come', would be the cry when floods devastated huge swathes of the plains and deltas.

The major river of the north, the Huang He, became known as 'China's Sorrow' due to the devastation it wrought within the historical core area of Chinese development (to be discussed again in its modern context in Chapter 7). Although in more recent centuries, flooding has become more associated with the 'Great River' (Chang Jiang, or Yangtze), it is nonetheless China's Sorrow that plays a greater part in the mythology and symbolism of flooding and the famine which regularly ensued, and in the actions taken to try to prevent it.

One of the earliest legends of China dates from the quasi-mythical Xia dynasty, the first dynasty, which according to official reckoning was founded around the twenty-second century BC and lasted until the sixteenth century BC. Before that date, however, there was according to myth a 'Golden Age' from 2357 to 2205 BC. But even this golden age suffered from floods. The then Emperor, Yao,

> called on his minister Yu to cope with the floods, which were devastating the country. This Yu did with great energy, fidelity and self-abnegation: 'Yu checked the overflowing waters. During thirteen years, whenever he passed before his home, not once did he cross the threshold. He used a chariot for land transport; a boat for water transport; a kind of basket for crossing marshes; and crampons for crossing the mountains. He followed the mountains in separating the nine provinces; he deepened the river beds; he fixed tribute according to soil capacity; he made the nine roads usable; he dammed the nine marshes and surveyed the nine mountains'.
>
> (Chavannes, 1898, cited in Tregear, 1980, p. 51)

This became the 'Legend of Yu', an exemplar not just of the best type of dedicated Chinese official, but also of the organisation and hard work required to ensure that floods are coped with adequately and their threat reduced. Yu tamed 'The He' ('The River') as it was known, by the following means:

- Canals were built to draw water from its raised bed after it entered the North China Plain;
- The river was split into nine rivers via channelling, before being allowed to return to a single channel further downstream, when its flow was slowed down;
- Dykes were built along the river banks. (ibid.)

Or, to put it more graphically:

> When widespread waters swelled to Heaven and serpents and dragons did harm, Yao sent Yu to control the waters and to drive out the serpents and dragons. The waters were controlled and flowed to the east. The serpents and dragons plunged to their places.
>
> (Schafer, 1967, p. 80)

Box 2.1 Li Ping: 'Dig the beds deep; keep the dykes low'

In Sichuan province, near the end of the second century BC, the first governor of Chengdu 'planned and executed one of the world's most remarkable hydraulic engineering achievements' (Tregear, 1980, p. 55). The plain of Chengdu was at the time a stony desert to the north and marshland to the south, criss-crossed by mountain torrents from the edge of the Red Basin to the west. Li Ping hit on the idea, similar to the legend of Yu (see main text), of splitting the channel of the main river, the Min, leaving one channel to flow in its old bed southward, but leading the other channel to the east. Subdivisions of each of those were then made, into numerous channels to form a network of irrigation channels. This was done using the 'arrowhead' principle of stone in midstream, supported by huge long 'sausages' of boulders bound into bamboo strips, each sausage being seven metres long and half a metre in diameter, on the arrowheads and along the banks. The channels are cleaned of debris annually, and the sausages are repaired or replaced by teams of people from the villages of the plain. This system 'has been properly maintained down through the centuries, and excellent crops of wheat, rice and other crops are produced' (Buck, 1973, p. 48).

Li Ping's maxim, which he had inscribed into the side of the gorge at Kwanhsien, where the Min had cut through the Azure Mountains to enter the plain, was 'Dig the beds deep; keep the dykes low', a principle which has been faithfully followed in Sichuan province for over two thousand years, thus avoiding the disasters of other areas where the beds became too shallow and the dykes were built too high. The initial work was long and arduous, so much so that it fell to Li Ping's son to complete the task. More than a thousand years later, in the Yuan dynasty, further channels were added, faced stone was used to line the dykes, these being cemented with lime and wood oil, and the banks were enhanced via large-scale planting with willows. In all, what we would now call a sustainable network of channels was built up, 'which irrigated, and indeed still irrigates the plain, turning it into probably the most densely populated, the most fertile and the most productive rural area in the world' (ibid.).

Yu was to become the founder of the Xia dynasty as 'Yu the Great', and although he 'was perhaps more perfect than probable, his labours are certainly a legend in which the drainage works of many generations have been assigned to one hero' (Fitzgerald, 1986 (1935), p. 27), he nonetheless symbolised what could be done to control the waters. For example, techniques of splitting the channels via 'arrowheads' of stones as well as other techniques became a feature of water control in Sichuan province, near the city of Chengdu. The indefatigable Li Ping led these developments, summarised in Box 2.1. However, as we have also seen in modern times, those in power must work ceaselessly to tame the rivers; when the powerful become too venal, lazy, cruel, corrupt or indifferent, the basic order in the system breaks down and disaster follows.

Then there is also the power of nature itself with which to contend. We can ask the question whether nature can indeed be 'tamed' or 'controlled'? Along coastlines, for instance, sea walls often displace erosion and deposition to another location. Dykes protect, but must be maintained regularly and consistently in order to do so, and once they are breached during a peak flow situation, the results may be more disastrous than if they had not been constructed in the first place.

Not only were dykes regularly breached by the Huang He (more than 1500 times in 3000 years) and other rivers; Huang He also regularly sought a new route across the North China Plain. It has been recorded that it made major changes to its channel no less than seven times, and 26 in all in the last 3000 years, migrating frequently from north to south of the Shandong peninsula and back again, a distance of hundreds of kilometres. Indeed, in 1938, in order to stem a different sort of threat, the tide of the invading Japanese army, the nationalist Guomindang government took the desperate step of cutting the southern bank of the Huang He via controlled explosions. This drastic action delayed the Japanese by a precious three months (Spence, 1990, p. 449), but resulted in the flooding of 54,000 sq. km of the Huai basin to the south of Shandong, including 4000 villages, and is estimated to have caused the deaths of nearly 900,000 people, plus the return of the Huang He to its southern course once more. This latter situation was not reversed until 1947, via international intervention (Tregear, 1980, p. 273).

The Huang He is known colloquially as the 'Yellow River'. This is due to the voluminous quantities of silt that the river picks up in its journey through the loess lands further west, before carrying this slowly and tediously across the North China Plain. 'It is no exaggeration to say that it leaves the gorges at Sanmen like a thick, yellow soup. The Chinese have a saying that 'If you fall into the Hwang-ho [sic] you never get clean again' (ibid., p. 274). The river consistently deposits this soup in its bed, thus raising itself steadily upwards, above the level of the plain itself:

> Of the yearly 1.6 billion tons of sediment discharged into the North China Plain by the lower Yellow River, about one-fourth is deposited on the river bed, which is being raised at a rate of 10 cm each year.
>
> (Zhao, 1994, p. 150)

Although the dykes are regularly built higher to try to deal with this, perhaps wrongly so (see Box 2.1), and the people are mobilised for extra work during the danger season, not only does the height of the river further endanger the people of the plain when it does burst its banks, it also means that the river cannot return to its previous course once the flood is ended, hence encouraging the changes of channel noted in the previous paragraph. One solution to this perennial problem was devised by hydrologist Wang Jing in AD 70, when he had a new channel dug, similar to part of the legend of Yu noted above, due eastwards to the Bohai Sea, from Kaifeng. This solution worked for nearly 1000 years. Another solution to the problem of silting was to narrow the river channel through building the dykes further into the river, thus speeding up the flow and reducing the silt deposition. This was led by engineer Pan Xiushun during Ming times, and proved to be a 'good tradition' for later centuries (ibid., p. 151).

Another knock-on effect of this flooding is the effect on other neighbouring river systems, especially on the Huai He. Because of the Huang He's regular invasion of this area:

> the original Huai he river system has been badly deranged (sic) and drainage has been greatly impeded. The result is, as the saying goes, 'Light rains cause light waterlogging; heavy rains cause heavy floods; while no rain results in drought.
>
> (Ibid., p. 205)

It is not just floods (which, after all, do deposit useful silt over the plain, potentially aiding longer-term agricultural output after the severe short-term problems are dealt with) that are a problem in this area, therefore, but also, somewhat paradoxically, the perennial threat of drought (see Box 2.2). The rainfall regime in North China generally is erratic. As we shall see later, when available rainfall is combined with over-extraction of water for the burgeoning industries and cities of the modern era, the threat of drought becomes especially severe. Salinisation is another problem in the area. Solutions to such issues include dams, but these too, as with traditional dykes, can be problematic solutions to control of river flow. Perhaps, in the longer term, the river dragon needs to be mollified rather than chained. This will be examined in Chapters 6 and 7, when we consider some of the modern solutions to reducing not just China's Sorrow, but more generally, China's other sorrows caused by too much water, or too little.

Han expansion – desertification and deforestation

The heartland of the Han (Chinese) people is, as we have seen, located in North China, in the valleys of the loess plateau centred on the Wei He/Huang He confluence and such ancient cities as Xianyan, Xi'an and Luoyang (see Box 2.3 for a discussion of the relationship between cities and the countryside). Despite the traditional image of rural China dotted with peasants toiling in paddy fields, the climate here is humid or sub-humid warm temperate and the main crops include winter wheat, corn and millet. Rice

Box 2.2 Droughts

Droughts are another severe problem, with 1056 of them recorded from 206 BC to the founding of the PRC in 1949 (Zhao, 1994, pp. 148–9). The early seventeenth century was a particularly harsh period – both a symptom and a cause of the breakdown of the Ming dynasty. There were 16 years of drought, for example, from 1628 to 1644, especially in North China, influenced by what has been called the 'Little Ice Age' across the globe, as well as epidemics and the abandonment of major irrigation and flood control projects. Millions died and the Ming Dynasty collapsed.

In the late nineteenth century, another drought, perhaps the worst ever known in terms of lives lost, hit the four provinces of Shanxi, Shaanxi, Henan and Hebei from 1876–9, killing 9–13 million via hunger, disease or violent upheaval (Tuan, 1970, op. cit., p. 165). Another disastrous period was 1920–1, when famine caused by droughts in 1919 hit most of Hebei province and surrounding provinces of Shandong, Henan, Shanxi and Shaanxi. Spence (1990) vividly describes the horrors that ensued:

> . . . the combination of withered crops and inadequate government relief was disastrous: at least 500,000 people died, and out of an estimated 48.8 million in these five provinces, over 19.8 million were declared destitute. Houses were stripped of doors and beams so that the wood could be sold or burnt for warmth; refugees crowded the road and railway lines, and many lost limbs or were killed trying to force their ways onto overcrowded trains; tens of thousands of children were sold as servants, or, in the case of girls, as prostitutes and secondary wives. Epidemics – typhus being the most dreaded and the most prevalent – dominated those already too weak to fight back (p. 309).

This horrific situation, and others like it during the inter-war years, helped create the preconditions for the Communist Party of China to flourish, and eventually led to the establishment of the PRC in 1949.

can, however, be grown on the North China Plain during the hot summer months. From this heartland, the Han people have expanded through time and over space to predominate in the 'eighteen provinces' of 'China proper', i.e. the area to the east of the upland regions, abutting the coast. They also occupy many other areas beyond, especially the cities. Some of this expansion was on a voluntary basis, especially towards the fertile Yangtze area, but some was precipitated by official decree and some was a reaction to

Box 2.3 Ecology and the ancient city

In the West, the symbiosis between city and countryside has largely been forgotten, whereas 'the close interdependence of city and countryside was far more explicitly recognized, and indeed welcomed, in China than elsewhere' (Murphy, 1980, p. 22). China's ancient cities grew with, rather than on, the surrounding countryside, and rural life was likely to be admired rather than denigrated in Chinese literature. A seventeenth-century official was not being untypical in noting that 'Goodness develops only in the village, evil in the city' (ibid., p. 24).

Ancient cities were laid out according to principles of Chinese cosmology, with geomancers employed to ensure good *feng shui*, in a very regular pattern, varied only by topography and occasional whim of those in power. Within the city, the temple of the earth, devoted to the god of the soil, became a key feature more than 2,000 years ago, as did the enclosure of agricultural land within the city walls: '. . . the vast majority of city dwellers, not only in Chou times but even in Han and later periods, were cultivators who, in summer at any rate, went out daily through the city gates to work in their fields' (Wheatley, 1971, p. 178). With them would go the carts laden with night soil (human waste) to fertilise these fields.

Urban ceremonials were much devoted to the rhythms of the agricultural season: ' the coming of the rains at appropriate times, the fructification of the earth, the springing up of crops, and the maintenance of the parallelism between nature and society which subsequently became symbolised in the harmony of *yin* and *yang*' (ibid., p. 479). This emphasis on harmony was important, and reflected the symbiotic rather than parasitic relationship between the city and its ecosystem. Of course, at times this harmony would break down – when urban officials became too rapacious, extracting too high taxes from the peasantry for example; when the city became too ostentatious and ornate, sucking up resources from far afield; or when disaster of famine and flood endangered the essential food supplies for the urban population. The huge banquets of Kublai Khan in his capital, Dadu, for instance, placed great pressure on grain supplies, to the extent that canals were dug southwards. Grain was also transported by sea from the rice bowl of the Yangtze valley, in attempts to ensure adequate supplies.

Water control, too, was an important element in urban–rural relations. In his study of the rise of Suzhou, for example, Marmé notes that without this, 'the area would remain, quite literally, a backwater. If sodden marshes were to become productive fields, dikes had to be built, and the waters behind them drained. The network of polders and canals which resulted was extremely delicate' (1993, p. 20), necessitating regular maintenance and periodic dredging of the delta's main river channels.

pressures from and invasions of the nomadic peoples to the north, or to the floods and famines discussed in the previous section. As they have migrated, so too have they brought, *inter alia*, the twin evils of desertification and deforestation with them. It is to these ancient legacies that we now turn.

It was under the legendarily brutal first emperor, Qin Shi Huang Di, that rapid spatial expansion of the Han people began, which then continued in the early years of the following Han dynasty around 2,000 years ago. By AD 2, the first census revealed a population of nearly 60 million, far beyond corresponding population centres around the globe at that time. This population, interestingly, is thought to have remained more or less static until the eighteenth century, although subsequent censuses until the late eighteenth century are reckoned to be underestimates due to what we might now call a 'poll tax effect', as people sought to avoid taxation. Nonetheless, even if the population at census dates was largely unchanged, great loss of life took place in North China between these dates due to the combined threats of the northern nomads, floods and famines, which together killed millions of people and caused great 'panics' that displaced large numbers southwards.

One reaction to the threat posed from the north was the building of the Great Wall. Originally a series of different walls built by different states, Qin Shi Huang Di unified these into the 'Great Wall', which ran 1400 miles from the sea in the east to the north-west frontier of the previous state of Qin. Apart from crossing political frontiers, this 'was also the first time that an ecological frontier was crossed on a wide front. The inclusion of the Ordos desert was the beginning of strategic advances into non-agrarian areas' (Wertheim, 1975, p. 40). Not only was this a vast undertaking, which itself caused the death of thousands of labourers – with a widespread view held today 'that a million men perished at the task, and every stone cost a human life' (Fitzgerald, op. cit., p. 140) – but its environmental impact was immense. The labourers working on the wall had to be fed and housed, and materials and supplies had to be transported to them via the new roads the Emperor ordered built. People were ordered or encouraged to move into what was a potentially unstable ecosystem, one which could, and would, be destabilised by human action.

Over ensuing centuries, Han Chinese settlers steadily migrated into Shaanxi, Inner Mongolia or further towards the north-west, into the grassland areas. Some were sent for mainly strategic reasons to secure the frontier, with huge numbers involved, including an estimated 700,000 into the Gansu Corridor during the reign of Han Wu-Ti (from 141 to 87 BC) (ibid.). The agricultural practices of the settlers proved inappropriate to the new conditions. Farmers were continually obliged to move on, for their 'ploughing had loosened the grassland soils, exposing the lower sandy layers which drifted under windy conditions' (Edmonds, 1994a, p. 36). Zhao notes that pre-existing pastoralism was appropriate to such areas as the Mu Us area to the south of the Ordos Plateau, for example, but around the city of Tongfan, which was founded in AD 413 and became a thriving city of 200,000 people in a fertile area, 'ever-increasing human activities caused the desertification process to begin', so that 500 years later the city had been completely overwhelmed by shifting sand. This whole area became 'Sandy Land', and is 'the most notorious example of the southward march of *shamo* (sandy desert) in China' (Zhao, 1994, op. cit., p. 158).

The upgrading and extension of the Great Wall by the Ming, plus intermittent warfare with the tribes to the north, contributed to destabilisation and degradation of the environment in many areas, encouraging further desertification. For example, the Ming built the Great Wall along the southern edge of the Mu Us Sandy Land, demarcating pastoral from crop-growing areas, but large patches of shifting sand began to appear within the latter, which were over-farmed. As is well known, traditional Chinese agriculture has an amazing intensity of land use, and is more akin to garden agriculture than field agriculture. Weeding is intense, and 'foreign' biomass is removed – and has been for millennia:

> They clear away the grass, the trees;
> Their ploughs open up the ground.
> In a thousand pairs they tug at weeds and roots.
> (Verse from the ancient *Shi Ching*,
> cited by Tuan, 1970, p. 61)

Use of night soil and other natural fertilisers (see the next section) helped keep such a system in balance, and it generally worked well in the fertile delta areas. On the unstable loess plateau or the Ordos Plateau, never mind the grasslands of Inner Mongolia or the margins of the desert areas themselves, however, the removal of biomass, especially of shrub or trees, created instability and stimulated desertification. Government policies would periodically exacerbate the situation by encouraging peasant migration into these frontier areas. The Qin, Han and Ming examples were noted above, largely via the Great Wall. In the mid-nineteenth century it was the Qing dynasty that encouraged the opening up of the Mongolian 'wasteland':

> As a result, large tracts of sandy lands in the south eastern Ordos Plateau were ruthlessly cultivated, which resulted in the devastation of grassland and the encroachment of shifting sand. Subsequently, the line dividing pastoral and farming areas moved northward to the present provincial

boundary between Shaanxi and Inner Mongolia, while along the Great Wall, a 60 km wide belt of shifting sands has been formed over a period of about 300 years (from middle Ming Dynasty to late Qing Dynasty).

(Zhao, 1994, op. cit., p. 159)

Unfortunately, as we shall see in Chapter 3, these lessons of history were not heeded; the modern era, too, has seen the exhortation of government combine with application of inappropriate techniques and methods to further destabilise these areas. Although, as we have noted elsewhere (Cook and Murray, 2001), desertification is not solely the product of human action for it also reflects cycles of climatic change, such human activity has contributed markedly to the spread of deserts, in north and north-west China in particular.

Desertification in these areas is closely linked to deforestation, albeit that the latter is more widespread. Deforestation is also said to have begun on a large scale more than 2000 years ago in the time of the Qin and then the Han dynasty. Forest cover may have been 40 per cent at one time, with 80 per cent in eastern monsoon China, 5 per cent in north-west arid China and 10 per cent of the Tibetan Frigid Plateau (Zhang, 1994, op. cit., pp. 161–2). Today, it is around one-third of that figure in total, far less than the estimated 30 per cent required for a stable ecosystem. The loess plateau became the classic example of the perils of deforestation, with the Yellow River becoming that colour due to the accelerated erosion of this area two millennia ago as forest cover was removed. In later centuries, the process of deforestation spread southwards to the Yangtze and beyond, as members of the old northern aristocracy led the exodus to this fertile warm region:

> Followed by pioneer farmers, they began to strip the subtropical hill-sides. Firewood was needed, and timber for buildings. The expanding bureaucracy needed unlimited supplies of carbon-black ink, taken from pinewood. Aromatic and medicinal barks were found in great abundance. Mercury and gold lay in wait under the soil, in amounts that must have dazzled the former inhabitants of the bare Yellow river valley. The rich wilderness was endless, and no one thought about its conservation.
>
> (Schafer, 1967, op. cit., p. 18)

The deforestation process accelerated from the time of the Ming, as large-scale urban construction took place:

> Starting from the Ming period (1368–1644), all the forests in the central part of Huang River Valley as well as Xiang River Valley were seriously denuded. The remaining forests in the mountains to the west of Beijing were exploited to build palaces at the beginning of the Ming when the dynasty established its capital there. As the forests around Beijing were

insufficient, timber for palace construction was shipped to the capital from as far away as Yunnan, Shanxi and Hunan. Firewood became so scarce that it had to be shipped to the capital from hundreds of kilometres away.

(Edmonds, 1994a, op. cit., p. 32)

Many of the great buildings of the Ming, such as the Temple of Heaven or the Forbidden City in Beijing, were made of wood; charcoal was used for the making of bricks or a range of other products such as iron or copper cash, and pine forests were harvested for ink, as noted above. So the forests were stripped, while agricultural activities also expanded in area and further into the uplands. As with many agriculturalists around the world, the forest was 'an enemy' to be cleared, so that crops could be planted and the wood used. Security would also be enhanced, as potential shelter for bandits or wild beasts would also be reduced. When such practices are added to the pressures of population numbers of recent centuries, we have a potent recipe for ecological disaster. By the early 1930s, the great ecologist, R.H. Tawney, concluded that Chinese agriculture:

was beset by two interlocking crises: an ecological one characterized by soil exhaustion and erosion, deforestation, floods, and the pressures of the huge population on scarce available resources; and a socio-economic one caused by exploitative land-tenure systems, abuses by moneylenders, poor communications, and primitive agricultural technology.

(Spence, 1990, p. 430)

In the light of such severe crises, it is perhaps little wonder that Mao's concepts of rural-based revolution found, unlike many of the farmers themselves, fertile soil in which to prosper.

Ecological balance

As the great historian of China, C.P. Fitzgerald, pointed out, peasant life on the North China Plain and the loess lands alike was particularly vulnerable to the vagaries of climate, especially rainfall:

The farmer's crops, and even his life itself, therefore depend on the spring rains, which are variable and sometimes fail entirely. If that happens the hot sun of the sudden oncoming summer withers the young shoots, and the crop is wholly lost. Should this calamity be avoided, there is the fear of an excessive rainfall in midsummer, which causes floods, washing away the unripened crops. In either case there is famine, and on the average one or other of these calamities afflicts the country or some part of it, every four or five years.

(1986 (1935)), op. cit., p. 35)

Without wishing to be thought environmental determinists, we concur with Fitzgerald's suggestion that such climatic conditions exercised a lasting influence on Chinese religious ideas, with prosperity dependent on a precarious balance of nature's perceived destructive powers that were capable, if uncontrolled, of inflicting immense catastrophes upon mankind. The earliest Chinese cults were therefore directed to maintaining, by magical forces, the harmonious balance of nature, which alone made possible man's life upon the earth (ibid., pp. 35–6).

In addition to the deities of heaven and earth (each having a temple in Beijing), many other lesser deities, too, were found, such as the god of the soil, or the grain god, for example, plus the spirits and dragons of the rivers and mountains. These could all be disturbed and made angry by human action, as with the building of the railways in the late nineteenth century. It was the foreigners who were blamed for this:

> Foreigners were accused of damaging the 'dragon's vein' (*lung-mai*) in the land when they constructed railways, and of letting out the 'precious breath' (*pao-ch'i*) of the mountains when they opened mines. The gentry held foreigners responsible for destroying the tranquillity of the land and interfering with the natural functioning of the 'wind and water' (*feng-shui*, geomancy), thus adversely affecting the harmony between men and nature.
>
> (Hsü, 1990, p. 390)

Such concepts have ancient roots and were largely Daoist (Taoist) notions, which among other outcomes directly contributed to the failed 'Boxer' rebellion that sought to rid China of the pernicious disturbing influence of the foreigners.

On the environmentally positive side, as we have noted elsewhere (Cook and Murray, 2001), the intensive agriculture that developed to support and stimulate the large population of China was often based on good ecological practices. The use of 'night soil' plus 'animal dung, pond and canal mud, oil cakes (refuse after pressing out the oil from oil seed crops), and some green manure crops' (Buck, 1973, op. cit., p. 64) all helped to return fertility to the soil, and pigs were used to recycle human waste. Waste, in whatever form, was minimised. A sixth-century AD agricultural encyclopaedia provides instruction for improving soil fertility, preventing erosion and improving yields, and protection from insects (Muldavin, 1997), while Edmonds (1994) traces such good practices even further back in time. Terraces, for example, were developed as a vital method for reducing run-off on steep hillsides, while the use of night soil was greater near to urban areas, thus easing to an extent the negative environmental impact of cities upon their hinterlands. Such practices are in harmony with nature, with humans being one with the world:

Oneness between Heaven and man [sic] can be also rendered as 'Heaven–man oneness' or 'nature–man oneness' according to the world order of the Chinese conception *tian ren he yi*. This key conception is a recurring thread throughout the development of Chinese thought. Its origin is usually traced back to Mencius (c.372–289 BC) or Dong Zhongshu (179–104 BC). I personally think that it can be dated back to Lao Zi (founder of Daoism) and even further back to *The Book of Changes* (*Yi Ching* or *I Ching*).

(Wang, 1998, p. 12)

In part, this oneness was almost literal, for the closeness to the earth was such that the Chinese 'are a people "whose dust was so intermingled in the soil"' (Payne, 1960, cited in Buchanan, 1970, p. 2). However, notwithstanding this intermingling, the environmentally positive legacies of the past were increasingly offset by the pressures exerted by the people as they sought to intensify agricultural production and extend the ecumene, or living area. With the introduction of what was, in effect, an interventionist Judaeo-Christian tradition via the perhaps unlikely routes of Marxism and Western science, and the abandonment (at least in official circles) of Daoism, which was seen as primitive or superstitious nonsense, typical of the feudal era that communism was sworn to defeat, the stage was set for the new environmental disasters of the PRC years, disasters that often had their roots in ignoring key lessons from the ancient legacies described in this chapter.

The new government, as we shall see in later chapters, often forgot the principle of 'take no action' (*wu-wei*). This does not mean doing nothing at all: 'Instead it advises a ruler not to take arbitrary, unreasonable or blind action, when it comes to governing the people or conducting state affairs' (ibid., p. 15). For after all, as Lao Zi said:

Nothing in the world is softer and weaker than water,
But in attacking the hard and strong, no force can compare with it,
For nothing can take its place.
The reason why the weak can overcome the strong and the soft can overcome the hard is known to all the people under Heaven,
But none would follow and practise it.

(Chapter 78, p. 99)

3   Politics in command

Among the principles on which the People's Republic of China, as epitomised by Mao Zedong, was founded, were those drawing on the Marxist materialist view that 'the earth could be controlled via human intervention' (Cook and Murray, 2001, p.198). Although the phrase 'politics in command' was developed to show command over the economy, it can also be extrapolated to command over the environment.

In this chapter we shall examine this in terms of 'heavy industry', the 'Four Pests Campaign' and related issues, 'the Great Leap and the communes', and 'tree felling versus planting'. The environmental legacies of these diverse policies have combined with those of ancient times already discussed to worsen the overall environmental situation.

Heavy industry

Traditional China was very much a rural society. True, cities were established from an early date, but were much interlinked with their rural hinterlands, and while such industrial activities as iron-smelting and brick production have ancient roots it was only with the Western intervention of the nineteenth and twentieth centuries that a meaningful level of industrialisation was introduced to China. Industrial products were not desired; the Chinese Emperor Qian Long stated to Lord Macartney when the latter led his trade mission to China in 1793: 'As your Ambassador can see for himself, we possess all things. I set no value on objects strange and ingenious, and have no use for your country's manufactures' (Schurmann and Schell (eds), 1967a, p. 102).

Such disdain contributed to Western pressures to open up China to the West's growing output of manufactures. Although the wars were ostensibly fought over opium, they were really about the right of the outside world to export to the presumed vast market of China.

On the Chinese side were many progressives who felt that China had to industrialise in order to compete with the might of the West, and although for many years such modernisers failed to gain power, the victory of the Communist Party of China (CPC) brought to the fore those who held to this view. Although leading a rural-based revolution, the CPC:

were determined from the outset to launch a grandiose programme of economic expansion. During the early years after 1949 the Soviet Union was their general model for industrialization and modernization, a model calling for rapid expansion of heavy industry with savings extracted largely from agriculture.

(Schurmann and Schell (eds), 1967c, p. 189)

Based on Soviet-style Five-Year Plans, heavy industry became the bedrock for Chinese economic growth for much of the 1950s–70s. In the 1950s especially, most opinion supports the view that China made rapid progress to recover from the destruction of World War Two and the Civil War. There were, for example, high levels of investment in the development of raw material extraction and processing industries; large-scale, capital-intensive technology in industry; high rates of saving and investment institutionalised through agricultural collectivisation (extracting a surplus from the peasantry) and an under-emphasis, by contrast, on agricultural investment, consumer industries and social 'overheads' (Eckstein, 1966, cited in Buchanan, 1970).

Steel, chemicals, coal, cement, and electricity production typified the production emphasis of this centrally planned economy (CPE), especially in the 1950s, but also in the decades of the 1960s and 1970s. Data became notoriously unreliable around the time of the Great Leap Forward (GLF) in 1958, but before that, fairly reliable figures show an increase from 1949–57 in coal production of just over four times, electricity production 4.5 times, steel production an amazing 334 times (from 0.16 million tons to 5.35 million tons), chemical fertiliser of more than 6 times, and cement more than 10 times (1949 data in Freeberne, 1971, p. 393–4; 1957 data in *China Statistical Yearbook*, 1995, pp. 412–14). As Li notes, 'For the first time and in quantity, the country was able to produce trucks and automobiles, merchant ships, tractors and jet airplanes, and to export whole sets of cotton textile machinery, sugar-refining machinery and papermaking machinery' (Li, 1967).

In addition, Mao Zedong, as a former guerrilla leader used to difficult times, 'wanted the granaries full and an iron and steel industry in every province'. In particular, 'His obsession with grain, the guarantor of life throughout the ages' meant an agricultural policy dominated by turning as much of the land as possible over to basic food crops. 'This meant trees being cut down, pastures being ploughed, grasslands destroyed . . .' (Pan, 1987, p. 22).

Christopher Howe, in reviewing the progress of the Chinese economy during the Maoist era, called the period from 1952–9, 'the Period of Fastest Growth' (Howe, 1978, p. xxiii). The financial resources for investment were *acquired* from agriculture and private industry, plus the profits of the expanded state industrial sector, while they were *allocated*, 'by establishing a Soviet style apparatus for the central planning of State industry, wholesale, retail, and foreign trade' (ibid., p. xxv). As regards capitalist enterprises, these were encouraged to merge with state enterprises as 'joint state–private operations', with the state investing in new equipment, often in larger factories, and gradually taking over the assets of the new operation, allowing the owners 'a fixed annual interest of five per cent of their shares in disregard of profit', an arrangement that was not terminated until 1967 during the Great Proletarian Cultural Revolution (Xue Muqiao, 1981, p. 30).

Successful though this period may have been in economic terms, with the modernising of old steelworks such as Anshan (Angang) in the Manchurian industrial area or Wuhan on the Yangtze, and the establishment of the major new plant at Baotou in Inner Mongolia, for example, environmental quality was very much ignored in the planning process. Cities became 'producer cities'. Even the capital, Beijing, was developed as an industrial centre: the Shijingshan iron and steel works (Shougang) in the suburbs was expanded and many inner-city industrial plants were allowed to grow unchecked. At the same time, urban development was rapid, whether in new industrial satellite towns as found around Shanghai or in older urban locations.

The quality of construction was often low, with bleak medium-rise blocks becoming a characteristic urban form in this era, albeit interspersed with superior buildings built in the mega-scale Soviet idiom, usually on broad boulevards. At the time, the goal was industrial development at all costs. Since then the cost of this rampant industrialisation has seemed far too high, not least the negative legacy of industrial pollution from the ageing plants themselves, together with the knock-on effects of industrial growth on the environment in general.

As regards Shanghai, one writer in the 1980s observed that the decay of its physical fabric, as well as that of other cities, was largely a legacy of the Maoist tendency to regard money spent on housing, heating, domestic electricity, transport, cultural facilities and so forth, as non-productive investment. 'Little wonder then, that China now has some of the most polluted cities on earth, and millions of urban Chinese have to live in overcrowded, squalid conditions' (Pan, 1987, op. cit., p. 51).

A major study by the Chinese Academy of Sciences of China's development trajectory, published in Chinese in 1989 and in English in 1992, noted that because the PRC came later to the development process, this gave the country some advantages, helping it 'to leap over several historical stages of development', but 'It has, however, also brought severe environmental pollution and ecological degradation at a lower level and stage of development' (Hu Anyang and Wang Yi, 1992, p. 38). In analysing the reasons for this, the investigating group firmly blamed the 'impetuosity' involved in setting economic strategies:

> In order to meet the ever expanding cultural and material demands of the people, we often commit the error of being impetuous in setting economic strategies, especially when we are under pressure from the outside world and are eager to shake off poverty. Our impetuosity is often reflected in excessive goals for economic growth and in overestimation of

our economic development and our ability to deal with difficult
problems and to tide over crises.

(Ibid., p. 62)

Of course, such impetuous behaviour is found in many countries too; the
difference in the PRC lies (a) in the fact that for many years, politics was firmly
in command at practically all levels of the economy, and (b) the sheer scale of
China multiplied to the nth degree such problems as poor decision making,
bureaucratic inertia and capricious leadership. The very nature of development can also be questioned; heavy industry may be a good foundation for
an economy, but is it a good foundation for a society? The share of national
product of industrial output, for example, was 28% in the early 1950s; by
the early 1970s it was 55%. The share for agricultural output was almost the
mirror image of this, from 55% in the early 1950s to under 30% in the early
1970s (Smil, 1993, p. 111). We can ask whether such a rapid transition was
quite so necessary, and whether the environmental price was too high to pay.

For example, coal production expanded from around 130 million tons in
1957 to more than 500 million tons in 1976, the year of Mao's death, and to
600 million tons in 1978 at the beginning of Deng's ascendancy to power
(ibid., p. 112). This coal was often of high sulphur content, especially from
the mines in Sichuan province, where it averages over four per cent sulphur,
and can reach nearly eleven per cent in some cases (ibid., p. 118). Coal
burning caused atmospheric pollution in the form of acid rain or smog,
stimulated respiratory problems and led to the less obvious, but difficult,
problem of ash disposal. As we shall see in the next chapter, these outcomes
have grown even more problematic since that period. Northern cities and
towns in winter especially can face a host of health problems brought on by
burning of coal for heat and energy production, and in Sichuan itself the acid
rain problems of Chongqing have become notorious (see Chapter 5, as well
as Cook and Murray, 2001). By the mid-1980s alone, 15,000 ha of pine trees
in Wan County, Sichuan, had already died due to acid rain pollution and
consequent increased vulnerability to pests and stresses, while another
38,000 ha were dying (Smil, 1993, op. cit., p. 120).

As for alternatives, oil production and petroleum refining became
environmentally problematic due to the problem of watercourse pollution
(see Table 3.1), with Lanzhou especially becoming infamous for the photo-
chemical smog caused by its petrochemicals industry combined with its
physical geography. Hydroelectric power seems a good alternative energy
source (see Chapter 8 for a discussion), but even that has problems, for
'Between 1957 and 1981, generation of every 100 GwH in China's large and
medium-sized hydroelectric stations claimed 50 ha of land and displacement
of 560 people' (ibid., p. 114), an issue that is magnified beyond measure for
the controversial Sanxia Dam (Chapter 6). It was the *intensity* of energy use
that was the main problem with the emphasis on heavy industry, however,
and although the graph for primary energy consumption continued to soar in

Table 3.1 Official estimates of harmful substances
discharged into wastewater, 1981 (tonnes)

Mercury	62
Cadmium	236
Chromates	2,367
Lead	3,006
Arsenic	1,200
Active Phenols	21,296
Cyanides	8,692
Oil	131,408

Source: Edmonds, R.L. (1994a), *Patterns of China's Lost Harmony:
A Survey of the Country's Environmental Degradation and Protection*,
London: Routledge, p. 134.

the 1980s and 1990s, the intensity of use declined since 1976, when the
emphasis was moved from heavy to light industry. This intensity of use was
also found with respect to water and other key inputs such as limestone while
the noxious and dangerous outputs included pollution of waste water by
mercury, cadmium, lead, arsenic and cyanides, and of the air by sulphur
dioxide and particulates. By 1981, there were 23,000 million tonnes of
industrial wastewater, of which only 26 per cent met effluent standards, and
just 13 per cent was treated (Edmonds, 1994a, p. 134). Table 3.1 shows the
main pollutants in wastewater in that year. In all, therefore, the emphasis
on heavy industry, as in the Soviet Union and Eastern Europe plus the
traditional industrial heartlands of Europe and North America (the
'Rustbelt', as it became known), led to severe environmental deterioration
and a negative legacy that has continued to the present day.

Four Pests Campaign

Apart from the ancient legacies of floods and famines, desertification, deforestation and drought discussed in the previous chapter, China has also had a
long list of pests with which to contend. In paddy fields, for instance, para-
sites are a threat to barefooted workers, for they can penetrate the skin and
thus enter the bloodstream. Floods carried snakes and other creatures with
them (probably the origin of the 'river dragons' discussed in the previous
chapter) and could spread pathogens, while the periodic plagues of the past
were spread by fleas from rats, various types of which became widespread. As
the Han people moved southwards so they began to meet the problems of
malaria, spread by mosquitoes, and of hookworms. The Tang poet Tu Fu
wrote:

South of the Chiang is the land of malarial pest,
Pursuing travellers without wane or gain.
(Cited in Buchanan, 1970, op. cit., p. 94)

By the 1930s it was estimated that two-thirds of the population of central China was affected by malaria, 'and throughout history it has undoubtedly been a factor explaining the aversion of the Han people to upland areas' (ibid., p. 94). The use of night soil, which we generally commended in the previous chapter, could be double-edged, and for example a 1920s campaign against debilitating hookworms, 'foundered on the age-old practice of fertilizing mulberry trees with human faeces, often infested with hookworm eggs' (Porter, 1997, p. 480). Typhus, too, was a common problem, especially in winter when it was spread by lice on peasants living cheek-by-jowl in crowded, squalid conditions. Other diseases were also widespread, often associated with squalor, poverty and poor sanitation.

The new regime targeted health improvements. But the dangers of 'the unreasonable application of reasonable principles' (Etienne, cited in Buchanan, 1970, op. cit., p. 204) in the 1950s were graphically illustrated via the 'Four Pests Campaign' of the 1950s against flies, rats, mosquitoes and sparrows. 'A rat is nine kilos of grain', 'Half the wheat and rice crops of Hunan and Kirin [Jilin] goes to fill the stomachs of sparrows', were some of the slogans employed to exhort the public to take action against these pests (ibid.). As we have noted elsewhere, 'Pots and pans were beaten continuously for days until the sparrows fell dead with exhaustion, and it was only some time later that the beneficial action of sparrows in the ecosystem was realised. Of course other bird species too would die' (Cook and Murray, 2001, op. cit., p. 200).

Bed-bugs were substituted in the campaign for the sparrows once the proverbial penny dropped, but by then the damage had been done, and the elimination of the sparrow was followed in some areas by a plague of the insect pests on which the sparrow and other birds feed.

Finally, another area of ignorance of ecological outcomes, as in other countries at the time, was the widespread use of DDT, which 'was being used both for agriculture and as a control of pests affecting humans' (Edmonds, 1994a, op. cit., p. 149). DDT has a number of environmental side-effects including biological amplification (it concentrates in organisms that cannot excrete it quickly as it passes along the food chain), which has lethal and sub-lethal effects such as egg-shell thinning and consequent lower rates of reproduction of those very predatory birds that could themselves keep insect pests in check; pest resistance, as evolution rapidly accommodates to such chemical threats and the potential replacement of one non-resistant pest by another resistant one; and hazard to farm workers in the areas sprayed as well as to the ecosystem itself (Simmons, 1996, pp. 274–5).

In China, pesticide use dominated the period 1958–70, and among other unforeseen outcomes, eliminated the natural enemies of the pest known as the plant hopper, in turn leading to a crisis in rice production between 1970–2 (Glaeser, cited in Edmonds, 1994a, op. cit., p. 149). In response to that, a move was made towards the much safer alternative of integrated pest management in the early 1970s, in which a range of methods and techniques are utilised, similar to some of those to be discussed in the next section. But we should also note that, in recent decades, 'with the rise of the reform movement and the return of the family farm, individual farmers have returned to the use of pesticides largely because integrated pest management is only successful if it is applied over a wide area' (ibid.).

To jump ahead of ourselves chronologically, we note that per-hectare pesticide use in grain production more than tripled during the last two decades of the twentieth century. During the 1990s, this resulted in many health problems. Indeed, deaths from the improper use of pesticides in crop production run about 300 to 500 per year. In order to get a clearer understanding of the consequences of pesticide use, a team of researchers from the Centre for Chinese Agricultural Policy at the Chinese Academy of Agricultural Sciences undertook an in-depth analysis in 2000 of the way the chemicals are used in rice production. The results were published in a report written by one of the centre's specialists, Huang Jikun.

The researchers studied rice production in Zhejiang province. Rice, which accounts for about 40 per cent of China's grain production, is the province's major crop. Although a lot of work has been done on the level of pesticide use in China, this was the first attempt to quantify its impacts on agricultural production and farmers' health.

To get the information they needed, the principal investigators worked closely with officials from the Ministry of Agriculture (MOA), provincial agricultural bureaus, crop plant protection stations, and county hospitals. Results from their field studies were cross-referenced with an extensive set of literature on agricultural chemical use across China. In Zhejiang, they found that the rate of pesticide use is more than double the national average. Average application of pesticide per hectare of rice (per season) amounted to 27.7 kg in dosage, or about 12 to 14 kg in active ingredients, similar to levels in Japan and the Republic of Korea, but much higher than in any other Asian country (Huang, 2001)

In light of this level of usage, the researchers set out to determine whether the quantity of chemicals they used actually helped farmers increase productivity. They found that, while pesticides contribute significantly to rice production, the benefit from increasing dosages declines considerably as the total amount of pesticide used climbs. In fact, they found that the benefits of using any additional chemicals were almost zero at current average usage levels. In general, the researchers estimated that farmers were over-using pesticides by more than 40 per cent.

To find out why farmers were using such a great amount of chemicals, the team investigated attitudes toward pesticide use. The researchers constructed a hypothetical behavioural model to see which factors were significantly affecting the behaviour of farmers. Analysis of this model allowed them to show that, not unexpectedly, it was the farmers' perceptions of yield loss that had the largest effect on how they applied pesticides. Farmers were found to grossly overestimate the crop loss due to pests: the average farmer's perception of yield losses was nearly twice the loss that actually occurred when no pest control was used. It was also clear that most farmers simply did not believe the recommendations and prescriptions on pesticide labels. Econometric analysis showed that education, farm size, occupation, and the quality of village-level extension systems were also all major determinants of how farmers perceived yield loss.

The researchers then looked at the health impact. The farmers they interviewed reported eye effects, headaches, skin problems, liver problems, and neurological effects, among others. Of 100 farmers examined, 22 had impaired liver function, while 23 had abnormal levels of key chemicals in their kidneys (Huang, op. cit).

Nationwide pesticide consumption averaged about 230,000 tons a year during the Ninth Five-Year Plan (1996–2000). Insecticides account for 70 per cent of all pesticide products, and medium to high-toxicity pesticides make up 70 per cent of all organic pesticides (*China Chemical Industry News*, 9 June 2001).

On Great Leaps and communes

To return to our historical review, one of the most amazing, and controversial, episodes in the life of the People's Republic was the Great Leap Forward (GLF) of 1958. The emphasis on heavy industry discussed above was at the expense of agriculture, from which the resources were squeezed to fund this industrialisation process. 'This neglect of agricultural development was reflected in agricultural output' (Li, 1967, op. cit., p. 197), with 1956 and 1957 being poor years, despite the increased collectivisation to these dates, collectivisation which the CPC generally, and Mao especially, felt had not gone far enough.

In the short term, the nation, but especially rural society, was mobilised with such exhortations as 'catching up with Britain in 15 years'. For the longer term, the communes were founded as the basic building block (along with state farms) upon which agricultural production especially would be based. As we have noted elsewhere, the communes

> were 'total' forms of social organisation which combined political, economic and social control over their area. They were continually modified in size, therefore their numbers varied from 23,630 in 1958 to 80,956 in 1963 to 51,478 in 1970 and 54,183 in 1980 (Chao, 1990, p. 131). As regards population, the communes contained over 560 million people in 1958, 568 million in 1963 after the famines of the 1959–62 period, up to nearly 700 million in 1970 and just under 811 million in 1980 (ibid, p. 131).

> (Cook and Murray, 2001, p. 199)

Table 3.2 Costs of a backyard furnace, Liu Ling commune, Shanxi

(a) *Inputs*	
Time	Four months (Oct. 1958–Jan. 1959)
Labour	67 men
Days' work	8,040
Wage costs	8,844 yuan
Iron ore	150 tons of 20 per cent iron ore, self-quarried
Coke	78.6 tons at 25 yuan per ton
Special bricks	340 yuan
Tools	498 yuan
Administration	143 yuan
Total Expenditure	11,790 yuan
(b) *Outputs*	
Iron	39.3 tons at 300 yuan per ton
Cost of iron to buy	289.40 yuan per ton
(c) *Profit/Loss*	Loss of 10.60 yuan per ton

Source: Extracted from the interview with Li Hsin-min, in Myrdal, J. (1967), *Report From a Chinese Village*, Harmondsworth, Middlesex: Penguin, pp. 445–6.

Most observers would argue that the GLF failed, that it 'was magnificent madness' as Schurmann and Schell put it (1967c, op. cit., p. 397). The hype was extraordinary, the effort prodigious. At first the results seemed amazingly successful; however, output figures were soon found to be exaggerated across the nation and this whole period has had to be revisited by China's State Statistical Bureau, with data 'retrofitted' downwards. Famines and flood were widespread, and famine took a severe hold in the bad weather years that followed. Figures of 25–30 million are now regularly cited as the death toll (taking into account children not born due to the knock-on effects of malnutrition of adults, as much as through actual deaths) for 1959–61.

One major campaign, the 'Backyard Furnace Campaign', in which 60 million peasants were exhorted to build mini-ironworks or steelworks and actually produced several million tons of iron or steel, is widely regarded as a failure due to the upheaval involved, time taken from more important tasks such as dyke repairs, and the poor quality of steel produced. Table 3.2 gives the final balance sheet reported by Jan Myrdal in his study of a village in the loess lands of Shanxi. The commune president who provided this information summed up 'that every ton of iron produced involved us in a loss of almost ten days' work. We decided, therefore, to stop the experiment' (Li Hsin-min, in Myrdal, 1967, p. 446). Li noted that the experiment did have some value, for people learnt the technique and iron had been difficult to buy at that time. Indeed, it has been argued that such exposure to industrial techniques assisted in the much later stage of the development of town and village enterprises (TVEs) from the 1980s, with some of these growing directly from those furnaces or other small rural industrial units that were deemed to be of higher quality and had better prospects. These had been allowed to continue after the furnace campaign was generally abandoned in 1959 (Li Choh-Ming, 1967, op. cit.), and are discussed in detail in the next chapter.

However, a positive feature of commune lifes, as we noted in our previous book, drawing on Buchanan (1970, op. cit.) was the 'Eight-Point Charter':

> in which agriculture was intensified via four activities which required more labour – water conservancy, close planting of crops, plant protection from locusts and other threats to crops, plus adequate manure via the use of 'night soil', green manure and chemical fertiliser – while those activities which needed less labour were better seed selection (of higher yielding strains), mechanization (usually semi-mechanization, what we would today term 'intermediate technology'), deep ploughing and better management.
>
> (Cook and Murray, 2001, op. cit., p. 000)

Irrigation and water conservancy projects were the key elements, and when the pace and scale of these projects were made more realistic in the decades following the mobilisation of the GLF, communes proved to be successful in building up 'communal capital' (Muldavin, 1997) via their work on this basic environmental infrastructure. Carrying on positive traditions of the past, millions of people at the grass roots level were regularly engaged in the winter months in digging out ponds and irrigation channels, weeding channels and ditches and freeing these from silt and other residues. The policy became 'three-pronged', the first being mainly small projects plus larger ones where feasible; the second, accumulation of water rather than its diversion; and the third, reliance on people rather than government. As Cook and Murray note, the policy was tailored to local conditions, 'thus there was an emphasis on drainage and water storage to fully utilise the water-logged lowlands near Tianjin, deep and shallow artesian wells were dug in the Beijing area as well as the loess land of Shanxi, while irrigation and storage was more important in the dry uplands of Hubei' (2001, op. cit., p. 199). Details of irrigation in one commune from a famous case study by Jan Myrdal are given in Box 3.1.

Communes were less than perfect in many ways, not least in stifling individual initiative, which retarded production, and so they were eventually abolished in the Dengist era. In their response to environmental conditions, however, as shown above they had much to commend them. Following their demise, the emphasis on making money produced an important surge in agricultural output and ensured that many people throughout the country, but especially in the cities, had access to good quality, varied and fresh produce in the food markets that quickly sprang up. However, the rural infrastructure has often been left to deteriorate in the process. For example, as noted in the previous section farmers have once again begun to overuse

Box 3.1 Irrigation projects in Liu Ling commune

One of the classic accounts of life in the commune era in China was by Jan Myrdal, who interviewed villagers in Liu Ling, in the loess hills of northern Shaanxi south of Yan'an in 1962. This village had been the location of one of the earliest experiments in collective and co-operative farming. The area 'is bare and barren, its climate hard. The crops fail there every other year, and its people are famed for their toughness, industry and rebellious nature. . . . The village lies 2,500 feet above sea level, on the loess plateau with their vast eroded slopes. Where the ground is not cultivated, the hills are covered with scrubby bush. The landscape is predominantly ochre in colour. Loess cannot easily absorb large quantities of rain. It can suck up a gentle rain, but here the rain is seldom gentle and so, when it comes, torrents of yellow muddy water gush down across the fields, digging deep furrows in the layers of loess. . . . Out of three years [the villagers] reckon on one good, one middling and one bad harvest. To them there is nothing unusual about crop failure; drought, frost and hail are a constant menace' (Myrdal, 1967, pp. 41–3).

For millennia, Myrdal's summary description had applied to this area, and perhaps to this very village. What was different by 1962, however, was that the villagers had been engaged in co-operative cultivation for 20 years, coping with such disasters as 'the worst flood in living memory' in 1942, as well as a Guomindang occupation in 1947–8 that destroyed 'everything', according to Li Yui-hua, the Old [Party] Secretary' (p. 431). The villagers rebuilt their co-operative endeavour through trial and error, until they found a system of book-keeping that fairly reflected the work that had been put in. They began preparing to be a commune in August 1958, allowing them 'to systematize joint effort' (p. 439) and thus making it possible to complete the dam that had been delayed previously because of the difficulties in getting the three agricultural co-operatives involved to collaborate effectively.

Liu Hsin-min, young ganbu [revolutionary; paid functionary], president of the commune and party secretary, stated that 'At the end of 1957, we had only 30 mu* of irrigated land. Since then, we have built 5 dams, 3 canals and 400 mu of terraced fields, and in 1962 the total irrigated land amounted to 8,320 mu' (p. 443).

Note: *A mu or mou is approximately one-sixth of an acre; one-fifteenth of a hectare.

chemicals to avoid the backbreaking tasks of weeding and use of organic fertilisers. This 'over-mining' of the communal capital and its implications will be discussed further in the following chapter.

Tree felling versus planting

Agriculturalists and 'developers' the world over have long denuded forests. This has given rise to various environmental concerns, including the impact on climatic change that has been expressed so forcefully in recent decades. As noted in the previous chapter, deforestation has a long history in China (see Boxes 3.2 and 3.3). The Han became so successful with this struggle with the forest 'enemy' that tree cover in some provinces – Jiangsu, Shandong and Hebei – became less than 0.5 per cent, and 'negligible' in the north-west provinces of Shanxi, Shaanxi and Gansu (Buchanan, 1970, op. cit., pp. 198–9). This lack contributed markedly to soil erosion and desertification.

When the CPC came to power in 1949, tree cover may have been as low as 5 per cent and was certainly less than 10 per cent. 'Make China Green' therefore became a rallying cry of the 1950s (as it is again at the time of writing – see Chapter 9), with mass plantings of shelter belts, forests of 'industrial', roadside and waterside trees. Official data showed afforestation to be 28.2 million acres in the First Five-Year Plan (1953–7), and in the GLF a further 69 million acres were claimed for 1958, plus 30,000 million trees around villages, along roads and riverbanks, followed by 39 million acres of plantings in 1959 (ibid.).

These figures, if true, would be amazing; even if considerably exaggerated, they would still be significant. Unfortunately, however, information released in the Dengist era revealed that the campaigns of the 1950s–1970s were not only grossly exaggerated but 'the real fate of afforestation campaigns disclosed a history of enormous fraud and waste . . . the stress on quantity and the neglect of quality resulted in a prodigious waste of labour and capital' (Smil, 1993, op. cit., p. 59). Smil comes to this conclusion because of fraudulent claims for the numbers involved, the low success rate for new plantings (below 30 per cent), and the relatively low level of tree cover per hectare compared to European forests. Even Buchanan, who praised the 'immensity of effort' being made to transform the vegetation cover, admitted 'the success achieved has not always matched the effort expended', and cites such reasons as:

- Inevitabilities, due to the marginal environments in which many plantings were made, including 'vast areas of unstable sands, erosion-prone mountain soils and soils with a very low nutrient status';
- Errors of judgement by officials, as identified by René Dumont, which included excessive emphasis on pines in timber planting and mistaken attempts to introduce fruit species into ecologically inappropriate environments;

- Absence of care after planting and wastage caused by faulty management techniques, with Richardson being cited as identifying a survival rate as low as 10 per cent (cf. Smil's 30 per cent noted above) for young trees;
- Lack of training in sylvicultural techniques and inadequate supervision of the planting process (Buchanan, 1970, op. cit., p. 201).

Trees can be planted relatively easily but their survival requires careful nurturing, and often this was not readily forthcoming because labour was directed to other activities instead. On the positive side, shelter belts were begun during this era, including one in Gansu Province in which more than 700,000 workers were involved, while later, from 1978, came the first phase of the 'Green Great Wall' (or 'Three Norths Shelter Project') across the northern provinces, designed to check advancing desertification, as noted in Chapter 1.

On the negative side, however, apart from the problems with afforestation noted here, the pressure on wooded areas continued unabated, with wood being required for fuel or construction, as well as still being removed to permit easier ploughing and cultivation. The backyard furnace campaign described above relied heavily in many areas on charcoal rather than coal, hence depletion in the forest cover was accelerated. Later, during the upheaval of the Great Proletarian Cultural Revolution, contradictory orders oscillated between encouraging the planting of trees and felling them for grain to be planted instead. Leeming (1985) cites one area in Anhui where fir trees were planted in 1964 prior to the Cultural Revolution, later removed for crops, replanted, removed, and then replanted once more in 1974 (p. 135)! In all, during the 1970s 700,000 hectares of woodland were cleared for arable farming, cutting against, as it were, the thrust of afforestation policies: 'For one man who picks up a spade to plant trees, there are a hundred who pick up axes to chop them down' (Tie Ying, cited in Leeming, p. 136).

Leeming provides considerable evidence of forest loss at the end of the Maoist period, gleaned from Chinese sources in the early post-Maoist period, now free to criticise. And so, for example, over-exploitation in mountains of the south had become a 'gross evil', with cutting in Fujian, Guangdong, Hunan, Anhui and Guizhou running at more than 10 million cu. m annually above production in each province, over 14 million cu. m in Yunnan, and cutting being more than 1.6 times growth in Sichuan (ibid.). Further:

> Tropical timber is a special problem. There were some 13 million mu of first-growth tropical forest in China in 1949; but in 1977, owing to excessive cutting and clearance mainly in Hainan island, only 3.6 million mu . . . in Hainan the whole environment is now under heavy pressure due to deforestation without proper protection or restoration of the deforested areas
>
> (Ibid.)

Box 3.2 Rise and fall of the state timber economy

The story of the state timber economy in China is one of collective exploitation of 'common' property. The central government has long overseen a state timber sector with 1.5 million workers, producing over half of the country's total timber. By 1992 accumulated state investment in 138 major logging companies was 11.62 billion yuan. These companies produced over 1 billion cubic meters of timber, paid taxes and profits of over 16 billion yuan, and accumulated fixed capital of 16 billion yuan (Li (ed.), 1996, Chapter 7, Natural Forest Conservation Action Program, 1997). Coupled with fixed low prices set on timber by the state, inadequate investment in replanting and inefficient management, the state timber economy finally drifted into a status of both financial crisis and resource exhaustion. The enormous forest stock inherited from before the PRC, plus a large workforce and sizable state investments, have not saved the state timber sector from deep financial and resource crises. As detailed below, harvestable resource stock in state company lands were close to depletion to 1995.

Harvestable forest area
1950 12 million ha; 1995 5.6 million ha. Rate of change: −53.0%

Harvestable forest stock
1950 2 billion cu.m.; 1995 0.47 billion cu.m. Rate of change: −76.6%
(Li (ed.), 1996, p. 163)

By the mid-1990s, over 80% of state forest bureaux were close to running out of harvestable forests, and some were already out of operation. On the financial account, many state logging companies were close to bankruptcy. Since 1990, the entire state forest industry has been loss-making, with an average debt/asset ratio of 70 per cent. About two-thirds of firms were overdue in wage payments and the unemployment rate had reached 17 per cent (Li (ed.), 1996, Chapter 7).

As a developing country, China has limited forest resources. According to the Fourth Inventory (1989–1993), the forest coverage rate was only 13.92 per cent, compared to a global average of about 22 per cent. By 1994, the total forest area in China was 134 million ha, with standing stock volume of 11.785 billion cubic meters. This total acreage of forests equal about four per cent of the world's total, whereas China's arable farming land and population are 7 and 22 per cent of the world's total, respectively. Among the existing forests, a large portion (34.3 million ha) are plantations. This makes China the country with the largest plantations in the world (Li, op. cit., Chapter 1;

FAO, 1997, Forestry Action Plan, 1995; UNDP, 1996), but also further points up the loss of original natural forests.

Some authorities have detected a change in government priorities over the half-century of Communist rule. This can be broken down as follows:

Year 1950–80
Timber/Fuel wood/Eco-protection/NTFP (non-timber forest products).

Year 1980–99
Eco-protection/Timber/Biodiversity/Recreation/Poverty relief/rural development/NTFP

(CIFOR, 1999)

Table 3.3 Annual Income Losses as a Result of Deforestation

Effects of deforestation	Income loss 1992 (billion yuan)
Reduced precipitation	81
Reduced lumber output	19.4
Desertification	18.81
Loss of water run-off	66.7
Reservoir/lake sedimentation	0.8
Siltation of previously navigable rivers	4.1
Property loss from flooding	13.4
Total	245.2

Source: Mao Yu-Shi, 'The Economic Cost of Environmental Degradation, A Summary', *Occasional Paper for the Project on Environmental Securities, State Capacity and Civil Violence*, World Resources Institute, 1996.

Bringing the story forward into the 1990s for a moment, Mao Yu-Shi has produced some interesting calculations on the economic costs of deforestation (see Table 3.3).

Leeming also gives further details of the problems of theft of timber (see also Box 3.3), which required the State Council in 1980 to issue an emergency circular to check indiscriminate tree cutting, of forest fires that are a perennial threat, and the ecological consequences of the clearance of the woods for arable farming, often in unsuitable areas, hence generating below average yields of grain and increased risk of erosion. In all, he attributes such problems to confusion of ownership and management rights to forests, plus waste and extravagant use of timber for construction or fuel, and, within state forests, conflict between different state organs, especially between the Forestry Department and the Agricultural Land Department, which have differing objectives, plus conflict between such bodies and local people (ibid., pp. 136–40).

Box 3.3 Forests lost to illegal logging

A warning that China's remaining 80 million ha of forest would disappear within ten years, if current logging methods were allowed to continue, was issued in 1998 by Jiang Zehui, President of the Chinese Academy of Forestry. 'History shows us that to protect forests is to protect our homes and the loss of forests is equal to the loss of our lands and lives', he declared (Deustche Press Argentur, 24 November 1998). The warning was issued shortly after China's worst floods since 1954, which officially claimed more than 3,656 lives and caused more than 300 billion yuan in economic losses. Subsequent investigation blamed the disaster on unrestrained logging along the upper reaches of the Yangtze River that caused serious erosion, raising water levels and silting up channels. The State Forestry Administration weighed in with the claim that a lack of forests was China's biggest natural resource problem. 'With a coverage of 13.9 per cent, forest shortage is even more severe than other resource-related problems, such as water, arable land and mining resources', it said. 'The shortage of forests has caused a series of economic, social and environmental problems to the country'. Green coverage per capita was only 11.7 per cent of the world average, with soil erosion covering 2.67 million sq. km – 38 per cent of the land (*Xinhua News Agency*, 2 September 1998).

The 1998 disaster – which some experts privately believed caused many more deaths than officially admitted – prompted the central government to ban natural forest logging with the hope of turning one million lumberjacks into tree-planters within a few years. Government officials also told the timber industry it had to abandon its traditional dependence on forestry and develop new materials – for example, using bamboo, sorghum and hemp to produce plywood. But any environmentalist tackling this issue quickly becomes aware of entrenched forces with a powerful vested economic interest in continuing unrestrained logging. In the three months immediately after the logging ban was imposed in 1998, local forestry security departments in Yunnan, Sichuan, Heilongjiang, and Qinghai provinces reported 26,369 violations, involving more than 28,000 people (*Agence France Presse*, 15 January 1999). Cheng Yu, a journalist with the Beijing *Guangming Daily*, a newspaper aimed at intellectuals, lost three fingers on his right hand when he was attacked by three men following the appearance of the last of his seven-part series concerning the illegal felling of more than 700 acres of virgin forest in Yu County, Shanxi province. Cheng had also uncovered evidence that this had the acquiescence, if not support, of the county's forestry officials. While officials presented a lumber licence dated from December 1998, Cheng quoted the villagers as questioning 'how the forestry department was able to obtain the licence when the central Government issued a ban on deforestation last year'. When the county forestry official changed his argument by pointing his finger at local villagers as 'lawbreakers for felling trees', the locals told Cheng they were actually 'the protectors' of the trees. Earlier in the year, hundreds of villagers detained a truck full of lumberjacks and carrying 71 logs. When the local forestry officials refused to take up the case, the defiant villagers travelled to Taiyuan, the provincial capital, to lodge a complaint. The county forestry officials were either dismissed or given an administrative warning (*SCMP* 15 October 1999).

In sum, therefore, the era of politics in command up to the death of Mao, despite some successes on the environmental front, generally left a legacy of severe ecological problems. In the next chapter we shall examine the subsequent era in which market forces were unleashed, to see whether their impact has been better, or worse, than this one.

4   Market forces unleashed

In the previous chapter, we considered how politically-motivated decisions aggravated environmental deterioration through the early decades of the PRC under the rule of Mao Zedong, when politics and ideology were very much in command. In this chapter we continue this discussion by considering the ecological implications of the industrial policies followed by Mao's successors.

These all had their roots in the decision taken by the late Deng Xiaoping to abandon the main tenet of Soviet-style communism, the centrally-planned economy, in favour of a system moving largely to the dictates of the market. Although it looked suspiciously like capitalism, an exploitative anathema the Maoist revolution had been designed to destroy, it would be termed a 'socialist market economy' with distinct Chinese characteristics.

In this chapter we look at the 'reform and opening up' programme launched by Deng in four key areas that impinge on the quality of the mainland environment: excesses in the practices designed to achieve greater agricultural production (referred to by some experts such as Joshua Muldavin as the 'overmining of communal capital'); the potential for creating pollution through the widespread establishment of Town and Village Enterprises (TVEs); importing pollution via the joint ventures that emerged from China's enthusiastic encouragement of foreign direct investment in a broad swathe of its industries; and exporting pollution from industrial development.

Pragmatism rules

In economic terms (although certainly not political ones) Deng Xiaoping was a pragmatist. His attempts to keep dogma out of economic management twice got him exiled from Beijing while Mao Zedong ruled, and could easily have cost him his life. But, once he was firmly in command at the end of the 1970s, he set out to undertake a massive reform programme that would modernise China and restore it to its former greatness of centuries before.

The concept was to 'unleash market forces', freeing the economy from many of the constraints of Marxist dogma and giving individual businessmen and women freedom from the dead hand of the central planner. But to assuage the unhappiness of hard-liners within the Communist Party that he was introducing once-vilified capitalism, Deng coined the expression 'market socialism'. The Twelfth National Congress of the Communist Party set out the goals of the 'socialist modernisation' in 1982. This called for a quadrupling of the gross annual value of industrial production by the end of the century, to place China in the front ranks of the countries of the world in terms of gross national income and the output of major industrial and agriculture products.

'Farms and factories, schools and service industries throughout the country [were] encouraged to 'put themselves in the market', to honour the principles of Adam Smith neatly refashioned in a socialist suit of clothes' (Murray, 1994, p. 8). Deng's stated goal was to prove that communism did not mean stagnation and inefficiency – both of which were rife in the state-run industrial system of the Maoist era – and that it could beat capitalism at its own game. This was to be achieved through the 'Four Modernisations': the upgrading of agriculture, industry, science, and technology. Intellectuals were encouraged to give free rein to their imaginations, with the expectation that great ideas would be turned into great products for the advancement of Chinese socialist civilisation.

The first step in agriculture and industry was to revive the 'individual economy', or private enterprise, that had been eliminated from the 1950s onwards, so that it could play a strong supportive role to the socialist (state) economy. Gradually through the 1980s, and more especially in the following decade, free markets were opened to allow peasants to sell their produce direct to the consumer at whatever price they could get (i.e. market-dictated) rather than selling everything to the state at an arbitrary rate. Factory managers were given more freedom to run their enterprises without constant interference from the in-house Party watchdog, including decisions on what to produce, where and how to sell the products and at what price, where and how to purchase the needed raw materials, and ultimately, decisions over the hiring and firing of workers (cf. Murray, 1998, Chapter 2, for a detailed discussion of this; also Wang, 1992, pp. 252–4; Evans, 1993, pp. 252–6).

In addition, power was increasingly devolved down from the centre to the provincial level and on further to the city, town and village government level in some cases (see Yang in Goodman and Segal, 1994, pp. 59–98). Thus local officials could decide industrial policy at the local level without necessarily having to have their decisions submitted up the chain of command to Beijing, with the resulting long delays and inefficiencies endemic in the bureaucracy.

In many respects, Deng's reform programme has been startlingly successful. China is, indeed, the third largest economic power behind the United States and Japan (cf. Chapter 1 of Murray, 1998, and Cook and Murray, 2001). But, as will be shown shortly, devolution led to local anarchy, with duplicated factory construction, production inefficiencies and heavy levels of local corporate debt, and insufficient attention to the environmental impact of industrialisation. Everything – logic, traditional local know-how, and good economic and business practice as taught on MBA programmes around the world – were all ignored in the stampede to unleash market forces at the local level.

Having once given local officials their head, the central government has found it hard to rein them in again. At various times, through the 1980s and 1990s, there would be abrupt U-turns in policy as the central government panicked at the loss of control and the discovery that allowing a large degree of economic freedom could not be divorced from the accompanying political freedom. Now, in the early twenty-first century, the central authorities find it increasingly difficult to bring provincial cadres to heel (raising grumbles in Beijing of 'economic warlordism' hearkening back to the 1920s when China was split up into various fiefs run by local warlords). And this, naturally, has important repercussions on the development of sound national environmental policies, as we will consider in due course.

Overmining of communal capital

For all its faults, such as the mad pursuit of ecologically unsound policies to constantly create record harvests during such periods of national myopia as the Great Leap Forward, Maoist agricultural policy still reflected to a significant extent the age-old wisdom of the peasantry in the growing of crops and the practice of husbandry. Efforts were made to keep decision-making at the local level in the belief that farmers acted in their own best self-interests to protect the soil and develop the local economy for long-term gain. Muldavin, for one, argues that this element has been eroded with the switch to Dengist reforms, to the detriment of the farming environment.

His research on 'the great northern granary' of Heilongjiang, where grain production is less than 100 years old, and single-season crop production only is feasible given the hard winters of the province, led to a critical analysis contrasting the favourable Maoist legacy of what he terms 'communal capital' with its recent 'mining' in the Dengist period, and the problems associated with the 'invisible hand' of the market. Due to such factors as intensification of land use and increasing size of herds, the outcomes include:

- Rapid decline in overall soil fertility in some counties (see Box 4.1 for a more recent example elsewhere in the country);
- Increased rate of organic matter decline in the reform period;
- 'Salinisation, ground water pollution, and micronutrient deficiencies' (Muldavin, 1997, p. 591);
- Short-term production gains but long-term problems of sodic alkalinisation;
- Increased patriarchal domination of decision-making;
- A riskier production environment requiring peasants to 'utilize practices they know undermine long-term sustainability' (Muldavin, 1996, p. 300);

- Increased disrepair in basic infrastructure of reservoirs, dikes, irrigation canals, and tree planting;
- Rapid increase in so-called 'natural disasters' at the local level as the protection afforded by tree-breaks, for example, is reduced, although there is also an element of seeking to win resources from state authorities by over-declaring such disasters;
- Inability to organise previously collective enterprises for, say, building a levee in Hesheng Village, leading to inappropriate yearly labour. The results are washed away by annual flooding, leading to demoralisation, eventual impoverishment of the village and enhanced out-migration.

From this, Muldavin argues that such issues are undermining the legitimacy of the state, especially at the local rather than central level, with rural unrest spreading in response to the wide range of problems associated with the reforms. He calls for 'locally-based collective action and long-term production strategies [which] must be promoted if the difficult problems of sustainability are to be resolved' (Muldavin, 1997, op. cit., p. 607).

One of the problems is that, although Chinese officials at all levels speak about 'sustainable' economic policies in both the agricultural and industrial sectors, it is often hard to know what they mean – if they mean anything. Central planning pre-1979 had its drawbacks, but at least it developed a vision for the future (or at least the ensuing five years) and everyone knew what their contribution had to be within the overall plan. Under Deng's 'opening-up' policies, referred to earlier, more and more responsibility for planning and execution has been devolved to lower levels, with accompanying lack of clear overall direction.

This was supposed to breed greater efficiency, in which the dictates of the market would be paramount, but has, instead, led to increasing 'short-termism'. The factory, the village, the city, county or provincial adminis-tration embarks on policies designed to maximise a revenue inflow, without considering where this might fit into the national picture and what possible negative side-effects might be produced for future generations. This aspect, within the terms of the definitions of 'sustainability' discussed in Chapter 1, should become clearer as we explore the issues in the following sections.

Township enterprises – rural driving force

One of the most important developments in the initial Dengist reform programme was the encouragement of 'township and village enterprises' (TVEs), small factories set up by local governments and local entrepreneurs involving various forms of ownership (with co-operatives the most popular form). These were able to recruit an average of 12.6 million workers each year from 1984 to 1988, but the numbers dropped sharply from 1989 to 1992 to an average of 2.6 million each year (Ministry of Labour reports for various years).

In the most developed areas of the countryside, by the mid-1990s the

Box 4.1 No more local liquorice

The humble liquorice root is one current example of the overmining cited by Muldavin. After lengthy study, the Academy of Sciences warned in early 2001 that liquorice in the mainland could be extinct within five years, without protective measures. Yu Jin, Vice Director of the Academy's Information Technology Centre, said one reason for such gloom was the fact that China had exported 1300 tons of liquorice roots in 2000, twice the 1999 amount, at a time when the basic resource was diminishing.

China had 2 million tons of root resources in the 1950s. During the same period, the total growing area in Ningxia autonomous region in the north-west reached 14.08 million mu (2.32 million acres). However, the acreage shrank to 8.8 million mu (1.45 million acres) in the 1980s with the uncultivated reserves of roots declining 50 per cent from the 1950s. In the 1990s, Ningxia's total growing area diminished sharply to 4 million mu (658,945 acres).

In Yanchi County, the main root-producing region in Ningxia, illegal harvesting damaged 3 million mu (494,209 acres) of reserves. Meanwhile, an average of 400,000 mu (65,895 acres) of grassland suffers deterioration and desertification every year. Yu noted that in other countries, the biological root resources were under tight restriction, while in China they were being exploited by villagers in pursuit of meagre profits. Liquorice is widely used in China in medicines and snacks. Disappearance would mean loss of valuable genetic information, Yu warned, in calling for immediate measures to protect the country's biogenetic resources. Overexploitation of biological resources not only caused the extinction of these resources but also destroyed the biological information related to them. He was supported by Liu Qian, Director of the China Biological Project Development Centre of the Science and Technology Institute, who predicted that ferocious international competition in the biological technologies and related fields would emerge in the coming years. Such competition would be characterized by a shift in focus from ownership of material resources to mastery of information resources, such as genetic information. (Author research)

TVEs were providing 80 per cent of household income and were seen by the central government as a vital element in dealing with the instability beginning to manifest itself due to mass unemployment in the countryside (see Murray, 1998, Chapter 3, and Cook and Murray, 2000, Chapter 10).

In the 1930s, sociologist Fei Xiaotong wrote a small book entitled *Peasant Life In China*, in which he suggested that running industry in backward rural areas might contribute to national economic development. In spite of his hopes, no genuine rural industry developed in the ensuing decades, until the Dengist reforms in the late 1970s provided the necessary impetus. However, the much-maligned 'Great Leap Forward' in the 1950s, discussed in the previous chapter, may have contributed some important groundwork. Western analysts have generally viewed it as a ghastly failure. But there is a Chinese counter-argument that while the steel produced in the backyard furnaces, in most cases, may have been of poor quality, the experience did, for the first time, introduce the peasantry to the rudiments of engineering and industrial production which would stand them in good stead in later years.

Whatever the truth, the fact is that township (formerly communal) enterprises have become the mainstay of the rural economy. They first appeared in a small number of villages and towns in the early 1950s, but due to lack of money, outdated technology and no access to information, they did not develop much until the family responsibility system was implemented in rural areas in the 1980s. That system quickly increased grain output, saturating the market and forcing farmers to seek other ways to bring in money.

From that point of view, they have been a success. Their value-added output in 1996, for example, reached 1,700 billion yuan, accounting for 60 per cent of rural added value output and 30 per cent of the country's GDP (*Beijing Review*, 17 November 1997). In the early 1980s, the government encouraged their development by setting low tax rates and exempting newcomers from any tax for the first three years. The special tax treatment was only abandoned in 1994, when a unified tax system was adopted for all enterprises. Cheap credit was also made available throughout the 1980s from the four specialised state banks. And although the preferential treatment has now generally been abandoned, it continues for enterprises located in the backward central and western areas.

But it is now apparent that they have reached a plateau and need to change direction. Jiang Yongtao, Director of the Agriculture Ministry's Township Enterprises Management Bureau, for example, told a symposium in Beijing that the township firms had grown by 17.3 per cent in 1998, contributing about three percentage points to the country's 7.8 per cent growth. However, their reform and development had entered a critical stage. Their exports had only edged up by 2.5 per cent in 1998 and more than 15 per cent of businesses were running at a loss as a result of the sluggish market and the impact of the Asian financial crisis, Jiang said. In the face of the difficulties, township firms should develop more in high-tech and environmental-friendly industries to find their way out, upgrading sub-standard equipment and training or employing more qualified people (Xinhua, 22 April 1999). But this will also need substantial investment by the national and local governments in raising the educational infrastructure.

Although generally optimistic about the sector, Chen Jianguang, a senior official in the Township Enterprise Department of the Agriculture Ministry also identified some shortcomings.

So far, few large companies have managed to emerge. Most continue to be small, poorly equipped firms, little different from the village workshops of several decades or even centuries ago. Typically, they comprise six or seven workers engaged in manual labour. The machines, if they have any, are old and usually second-hand equipment bought from state-owned enterprises. There are even some firms that are more like museums of equipment made in the nineteenth century.

(Interview, 12 August 1997)

In addition, the township enterprises have been censured for their contribution to a polluted environment.

For several years, the official media were full of reports about a government crackdown on notorious polluters, including some 5,000 small paper mills, tanneries and dye factories along the seriously contaminated Huai He, which were shut down by the authorities. And, early in 2000, Xie Zhenhua, the Minister of the State Environmental Protection Administration (SEPA), told a national conference on environmental protection that the government would continue to close small factories and mines failing to meet State pollutant-discharge standards, regardless of local interests (*China Daily*, 10 January 2000).

But this issue of 'local interests' is a sensitive one. For instance, the State Administration of the Metallurgical Industry has a plan to close down small steel smelters and rolling mills with a production capacity of less than 100,000 tons per year, as well as those that rely upon small and backward blast furnaces and steel converters, although no schedule for closure had been revealed at the time of writing. There are more than 2,500 small steel plants in the country, most of which are regarded as low-technology operations that waste resources and generate serious pollution. The closures are not just aimed at eliminating pollution sources, however, but also to curb the steel industry's chronic problem of over-production. For whatever reason, the elimination of such plants is unlikely to be welcomed locally and accomplished easily, as officials readily acknowledge.

Wang Xiaoqi, director of the Development and Planning Department under the Administration, conceded: 'It's not going to be easy [to achieve closures], nor will be it be without financial strain, but the government is resolved to carry out the plan because it will promote industrial restructuring and technical upgrading by phasing out backward and inefficient equipment. However, thousands of workers will be laid off, and in some areas that is going to mean real hardship.' Wang predicted some local governments would be unwilling to shut down the plants because they were major revenue sources, so that strict measures would have to be taken by the central government when and where necessary. For example, the plants to be closed would get no raw materials, no power supplies and no production or environmental protection licenses. Financial institutions would not provide the small plants with loans, and their steel products would not be granted market access. In addition, their equipment would be dismantled or scrapped to keep them from being reused elsewhere (*Business Weekly*, 9 January 2000).

Concern that the rapid development of TVEs would have enormous impact on future water quality was first raised in 1992 by Xu Fang and others, and in 1997 by Cao Fenzhong. According to Cao:

> Though their development can be traced back to the late 1950s, these enterprises boomed in the past ten years. The economic success of the TVEs has reduced poverty for millions of farmers, but they have also inflicted severe damage on the environment in rural China. Even though the government has enacted a number of laws and policies to control and regulate industrial discharges, [it] has not yet effectively regulated TVEs.
>
> (Cao, 1997, p. 1)

Further, by 1995,

> more than 7 million TVEs existed throughout China, with a total output of 5.126 trillion yuan (US$671 billion), accounting for 56 per cent of the total industrial GDP. The number of TVEs is expected to continue to grow. A conservative estimate holds that the TVEs discharge more than half of all industrial waste water in China – more than 10 billion metric tons. Most TVEs have no waste water or hazardous waste treatment facilities, and since [they] are widely scattered across vast rural areas, wastes from TVEs have the potential to affect the health of many people.
>
> (Ibid., p. 5)

Earlier, Xu had noted that a 1989–91 investigation of the ten leading TVE industries in seven provinces and municipalities showed that industrial wastes were discharged without any treatment and control.

> An analysis of the health of 860,000 people in the area revealed that the incidence rate of chronic diseases was between 12 and 29 per cent, much higher than the national average for rural areas, which is approximately 9 per cent. The total mortality in polluted areas averaged 4.7 per thousand, higher than the average 3.6 in the control area. Life expectancy in the polluted areas was two years lower than in the control area. Although not definitive, evidence suggests that industrial pollution from TVEs could become a major threat to human health in China.
>
> (Xu *et al.*, 1992, pp. 1–23)

One pollution accident that did cause health damage was the leak of waste water from the Chengzhou Arsenic Production Plant in Hunan Province on

8 January 2000, into the wells supplying drinking water to a nearby village. As a result, 255 villagers suffered arsenic poisoning, although none died. The plant was quickly closed down by the local authorities, its Vice-Director, Zhang Fangxuan, was arrested on a charge of criminal negligence, and all the village wells were sealed off to prevent further spread of the contaminant. According to local prosecutors, the factory had three ponds for containing polluted water, one of which cracked and caused the leak. But Zhang allegedly paid no attention to the split in the pond's wall, though factory staff told him about it more than once (*China Daily*, 17 February 2000).

Each year, large amounts of pollutants are dumped into water bodies from municipal, industrial, and agricultural sources, either by accident or by deliberate design. Except for some inland rivers and large reservoirs, water pollution trends have worsened in recent years, with the pollution adjacent to industrially developed cities and towns being particularly severe (Smil, 1996, p. 33). Some of the major threats to water quality stem from inadequate treatment of both municipal and industrial waste water. In 1995, a government report revealed the discharge of a total of 37.29 billion cubic tons of wastewater, *not including wastewater from TVEs* [our italics], into lakes, rivers, and reservoirs. 'Approximately 60 per cent was released from industrial sources, the rest from municipal. With only 77 per cent of industrial wastewater receiving any treatment in 1995, nearly one half of the industrial wastewater discharged failed to meet government standards.' (NEPA, 1997). Industrial discharges usually contain a range of toxic pollutants including petroleum, cyanide, arsenic, solvents, and heavy metals (*China Environmental Yearbook*, 1996, pp. 478–80).

Government environmental experts found that, although the amount of wastewater discharged from regulated industries had levelled off since the early 1990s, 'discharges from TVEs and municipal sources have increased rapidly' (NEPA, op. cit.). The increase from TVEs can partly be traced to the rising proportion of total industrial output from these enterprises, but we believe it is also due in large measure to a lack of pollution control over these enterprises for some years due to their widely scattered geographical distribution. In addition, local authorities have been reluctant to tighten control over pollution when pursuit of economic benefits is their first goal.

One could argue that the trouble with the TVEs is that they have been too successful. As with many aspects of life in the PRC over the past half-century, once the state media began trumpeting a new innovation – indicating, therefore, that it had the imprimatur of the central government and the Party – everyone wanted to be part of it. Thus, if cement became the flavour of the month, because the *People's Daily* had reported there was a need for increased output to meet construction demands, local authorities the length and breadth of the land would immediately issue orders for the construction of a cement plant. In this, one can see echoes of the Great Leap Forward's backyard steel furnaces discussed in the last chapter.

Thus, cement plants would proliferate, sometimes in sight of each other in different townships, with little consideration of an overall industrial plan nor of the need to consider quality and environmental protection. Box 4.2 provides an illustration of this from the perspective of one small city in southern China, and how the problem is finally being tackled.

Further north, the Yellow River faced a new threat in early 1999, when water in a 1000 kilometre-long section began turning turned brown and foamy. The situation became critical when the foamy water converged on the Xiaolangdi Water Control Project, under construction in Henan Province. As a result, a waterworks system in the city of Henan's Xinxiang City was forced to close down, while the water quality in Hua Yuankou, a major source of water for Zhengzhou, the provincial capital, was found to have fallen below state standards for drinking water.

'This is the most serious pollution ever in the history of the Yellow River', Cao Han, director of the general office of the water resources protection bureau under the Yellow River Water Resources Commission, told the official *China Daily*. 'The pollution has already affected an estimated 1000-km-long section of the river between Tongguan in Shaanxi Province and the river mouth in Shandong Province or about one-fifth of the total'.

Analysis and tests on the polluted water indicated that both chemical oxygen and biological oxygen demands (COD and BOD), two of the major water pollution indicators, greatly exceeded state standards, Cao explained. The causes were unclear, although the seasonal decreasing flow of the river and a reduction in the sandy content of the water that performs an absorption function were thought to be contributory factors. However, Cao believed the pollution resulted from the return of numerous small pollution-prone enterprises along the river or its tributaries, including pulp mills and leather-processing factories, and his agency quickly sent out notices urging local authorities to tackle the problem immediately. (*China Daily*, 29 January 1999)

Wang Xinfang, SEPA Vice-Minister told delegates to the Third National Work Conference on Environmental Protection for Construction Projects in Beijing in late 1999, that in regard to the 'Three Simultaneous Requirements (TSR)' – in which industrial waste treatment facilities are designed, constructed and operated coincidentally with the projects – there was still some way to go. Approximately 79,000 construction projects were initiated annually between 1995 and 1997, but only 80 per cent had met requirements. Between 30 per cent and 40 per cent of rural enterprises and 60 per cent of foreign-funded firms, in some provinces and municipalities, met the requirements, said Wang, adding: 'It shows our awareness of environmental protection is still low.' New projects in China were generating 800 million tons to 1 billion tons of waste water, 300,000 tons to 400,000 tons of sulphur dioxide and an increasing amount of soot annually, while environmental-protection investments in 1998 averaged only 5.3 per cent of money injected into construction projects, he pointed out.

Box 4.2 A case study of over-capacity

Zengcheng City claims one of the mainland's highest concentrations of cement factories and arguably Guangdong Province's most polluted air. As of the end of 1998, the satellite community, 60 km south of the centre of Guangzhou, was home to 110 cement-makers – observable to rail travellers by the red-brick kilns belching plumes of smoke and cement dust trackside and across the city of 750,000 people. Cement also dominates Xiancun, a suburb of Guangzhou itself. Clusters of belching cement factory smokestacks jut out from the hillsides, cement dust clogs the air and shrouds nearby fields.

According to Yuan Julun, deputy general manager of the largest producer in Xiancun, Zengcheng Cement, the local cement trade turned from a sideline enterprise into a village obsession at the start of the 1990s, when scores of smaller factories opened to help feed the frantic building and road construction. 'Demand was far greater than supply'. Mr Yuan said.' At that time even small factories were making 100 yuan profit on each tonne of cement they produced.'

However, as the regional economy subsided, vast over-capacity sent the entire sector into recession. State-owned Zengcheng Cement, with an annual capacity of 300,000 tonnes, has not reported a profit since 1995. As a result, management was forced to lease a company mine and a subsidiary cement factory, and retrench some of its labour force. Similar difficulties are reported throughout Guangdong province and the mainland at large. A survey conducted by the Guangdong Construction Committee concluded over-investment and repetitive construction were driving the entire industry into ruin, with as much as 67 per cent of production capacity in places such as Maoming City standing idle.

Since January 1999, however, a total of 40 Zengcheng cement producers have been closed, while the majority of the 48 producers in Xiancun suffered the same fate. The decision to close factories in Xiancun and Zengcheng is part of the central government's larger plans for restoring profitability to a catalogue of industries that are suffering from over-capacity, low efficiency and the use of outdated technology. This is not just designed to stop over-production, but also to shut down some of the country's worst environmental offenders. Ye Zhirong, of the Guangdong Environmental Protection Bureau, said it would close all Zengcheng cement factories producing less than 44,000 tonnes annually by the end of 2000. The State Economic and Trade Commission (SETC) targeted 10 manufacturing sectors, with 114 products, for immediate action in early 1999. Included on the list were coal mining, steel making, textiles and power production.

Guangdong's construction committee now grants production licences only to those cement manufacturers who are qualified under national guidelines, with the aim of shedding 100 million tonnes of national production capacity immediately. Moreover, stricter environment-protection regulations will force cement companies to raise production standards. According to Hao Zhenhua, an official of the State Administration of Building Materials Industry, the government would be closing 2000 illegal cement plants, shutting down another 676 firms whose licences have been revoked and close 490 production lines by the end of 1999. At the same time, it planned to close a number of small glass mills that annually produce no more than 200,000 boxes of glass each. Like cement, the glass industry has developed rapidly under the reform and opening policies. The output of flat glass increased from 17.84 million boxes in 1979 to 175 million boxes in 1998. At that time, China had 330 flat glass mills with about 500 production lines with a capacity of 215 million boxes. 'Small cement and glass factories play a definite role in local economic development, but their backward technologies, large energy consumption, low productivity, poor quality and their propensity to pollute resulted in serious negative effects to sustainable economic development', Hao contended. (Author interview)

Contribution of urban industries

But it is not just the township enterprises that are to blame. Even within the boundaries of the major cities, uncontrolled industrial expansion, especially of small, low-tech, TVE-style factories have also had an adverse impact on the quality of the air and water.

A case in point is the Suzhou Creek, which flows through the centre of Shanghai into the Huangpu River, and was in the early part of this century lined with parks for the exclusive perambulations of the Westerners ruling the roost in the international settlements that dominated Shanghai's economic and political life. Suzhou Creek used to be called Wusong Jiang (River) – the mother river for Shanghai – and the Huangpu was actually one of its branches. Its whole length is 125 km, of which 53 km is in Shanghai with 23.8 km snaking through the urban area. Deterioration began with the invasion of the Japanese in the 1930s. The parks disappeared, to be replaced by slum areas and industrial sweatshops. In the post-war period, heavy, highly polluting industry set up shop on the riverbanks, which became a sacrificial pawn in the industrial revolution. Year after year the river flowed by increasingly glazed with black dirt and spreading a choking smell that old-time residents have never forgotten.

Box 4.3 A model for clean-up

Of the six major downstream branches of Suzhou Creek, the worst was Zhengru Port, with water quality three or four times worse than the fifth category of water pollution. To tackle the problem, the Shanghai Fuxin Riverbed Treatment Co. Ltd cut off the water and diverted the stream into a canal where it could be treated. The company used pipes to suck away the black water and then put eco-friendly composite bacteria into the equipment, followed by the introduction of oxygen to help bacteria grow. The latter eventually ate up the organic substances in the dirt. The river water with bacteria was inserted into the silt in a pool one metre wide, 1.65 metre deep and 60 metres long, divided into six squares. The water flows from one pool to the next. In the first pool, the water is still black and smelly, but by third pool, after 30 metres, there is no more smell and the water is clear. By the sixth pool, it is clear to the bottom and there are even fish swimming in it; the water has reached the third category of the government's standard for earth surface water that can be used for irrigation and as raw water for tap use.

The company's general manager, Zhang Yongtai, used to be the chief representative of two Japanese companies in China, and also head of a trading company. However, he resigned to spend all his savings – 4 million yuan – on the polluted water treatment of Suzhou. Some say he is crazy to throw his money into river, but he says he is not someone who follows the correct line. 'I am not a fool', he insists. The environmental protection industry is full of opportunities, and treating the Suzhou River brings both social and economic benefits. Yang Xianzhi, head of the Shanghai Railway University Environment Health and Technology Research Centre, is a well-known environmental expert who has been researching Suzhou Creek pollution ever since 1968. But he felt his research was rather 'soft science' and he wanted to get into the 'hard side' of the environmental protection movement. The turning point came when he met Zhang, who sponsored him to go abroad to study water treatment water techniques.

Ever since 1960, Shanghai researchers had been working on how to treat pollution. They tried various diversion and water cleaning techniques, even using high-pressure gas jets to scour the riverbed. But, every time, the results were only superficial – surface cleansing only – and failed to tackle the underlying contaminants. In addition, they only tackled the creek's main trunk, ignoring the fact that dirty water was pouring in from many side streams, so that the overall pollution got worse.

Shanghai Fuxin is using the principle of one substance from nature to treat another. The bacteria to eat the dirt will not cause secondary pollution, but will create good circulation of river water and the balance of nature, the company says. And the process is also economical, costing 1000 yuan to treat one ton of water, whereas previous methods cost 2000–3000 yuan. According to forecasts, there will be 20 billion yuan invested in treatment of Suzhou Creek. According to the city's long-term environmental clean-up plan, 6.05 billion yuan was to have been invested in the treatment of branch river water, but with the Shanghai Fuxin technology the cost could be reduced to 4.76 billion yuan. From the model water treatment centre, the technology will undoubtedly be spread. (*Shanghai Economic News*, 4 October 1998)

In 1985, a Shanghai report forecasting the pollution level in the city in 2000 analysed environmental conditions around the Suzhou Creek area, and its potential damage to Shanghai. It warned that, every day, Shanghai was disposing 2.18m tons of dirt into the river, making the water quality worse than fifth category standard (seriously polluted degree). If left unattended, the cancer rate in Shanghai, and also the death rate, would be number one in China; food pollution would also be number one. This report was such a shock that, from end of the 1980s, the Shanghai municipal government has been putting great effort into a clean-up.

Mayor Xu Kuandi ordered a comprehensive treatment programme that would focus on riverbed cleaning and tree and other vegetation planting, which would hopefully turn the banks into something resembling the River Seine in Paris. The short-term target for the river clean-up by the year 2000 was to get rid of the black dirt and bad smell. There also had to be some obvious improvement in the cleaning up the dirt and the temporary installations put up by traders.

In recent years, some of the polluting factories were relocated to the surrounding rural areas, leading to the belief that this would effectively deal with one of the most important sources of pollution. But, there was no obvious improvement in the water quality – because the pollution was coming, and continues to come, from multiple sources. Besides industrial pollution, there is human rubbish and also rural and urban husbandry pollution together with the silt on the riverbed. Besides the main trunk, there are 35 branches, which have been developed as small ports handling the barges that shuttle between the big ships moored out in the Huangpu and beyond. They bring in black dirty water from a much wider region every day. The speed of polluting is faster than the speed of treatment. But some models for effective treatment are emerging, as discussed in Box 4.3.

Shanghai has relocated over 700 industrial enterprises from its downtown area to outlying suburbs, many of them textile and chemical operations. But

Box 4.4 Tale of two cleaner cities

Dalian, the key north-eastern port, is another city that suffered badly through the Communist penchant for heavy industry. But it too is now getting rid of the old polluting factories. For example, in December 1999, the remaining four buildings of the Dalian Dyestuff Factory were detonated into oblivion, creating a 550,000 sq. m area that will be used to build residential quarters, a school, a people's square, and lawns along the east beach of the city. In 1995, the city designated 134 factories, mostly state-owned enterprises, for removal, of which 85 had disappeared by the end of the decade, producing 2.4 million sq. m for redevelopment. The city's strategy is to eliminate pollution within the downtown area and rebuild it into a business, finance, tourism, and information centre for north-east Asia, said Lian Dong, an official with the municipal economic commission. (Author interview)

The Square of Haizhiyun ('rhythm of the sea') a favourite for citizens in the summer of 1999, was once part of the Dalian Dyestuffs production plant. The company operations once gave off terrible smells that forced its neighbours to close their windows even on the hottest days. The largest small goods wholesale market, the biggest sewage treatment plant and the highest-grade residential district of the city are also the results of factory removal and elimination of pollutants in Dalian.

Out of the limelight and with no recourse to outside funding, Chengdu municipality, meanwhile, is halfway through an ambitious project to move 200,000 people out of the city centre and clean up the two rivers they used as an open sewer. Unlike other resettlement projects in Beijing and the Three Gorges Dam area, the residents appear happy to be moving to new housing in the suburbs of the Sichuan capital. 'This project is not like resettlement projects in Beijing, where people don't want to leave', said Tian Jun, information officer for the Fu-Nan Rivers Project. 'We have so far moved 100,000 people into new houses. Their old accommodation on the river used to be flooded every summer and the conditions were cramped and unhealthy'. The project, which has so far cost 2.7 billion yuan, started in 1994 with the relocation of 650 seriously polluting factories. The 24.5 sq. km. of land reclaimed so far is being transformed into parks, including the world's first educational water garden, where water from the rivers moves slowly through a series of purification pools, said Tian. (Author interview) there are suggestions of a 'Nimby' (Not In My Backyard) mentality at work here. The enterprises, all designated 'serious polluters', were mostly moved to agricultural areas, where peasants hold less political clout than their urban neighbours and have less-developed regulatory schemes for protecting the environment, an official with the Shanghai Environmental Protection Bureau once admitted to *Xinhua News Agency*.

The official said relocation campaigns were common to the development of many coastal cities, and allowed industrial polluters access to cheaper land with no government interference. 'Local unprofitable firms see this as a good opportunity for future development', *Xinhua* explained. Shanghai plans to relocate another 2,000 industrial polluters to the suburbs by the year 2010, although some may be allowed to remain in the city centre if they convert to service-industry enterprises (*Business Weekly*, 13 November 99).

The experiences of two other widely separated cities, Dalian and Chengdu, are briefly dealt with in Box 4.4.

Role of foreign investment

Foreign direct investment (FDI) has generally been viewed as a good thing by the mainland, which has assiduously wooed investors with the profitable prospects of its huge market. FDI has been painted as a vital tool for China to modernise its industry and management methods. The hope, and general assumption, has been that foreign firms will introduce their latest technology, which will generally be of an environmentally friendly nature. By the late 1990s, at least 140,000 foreign-invested companies were operating on the mainland, with the vast majority causing the environmental authorities little concern.

But there have at times been accusations that investors are dumping their unwanted polluting industries on China (for an overview of the impact of FDI on China, see Cook and Wang, 1998, and Murray, 1996). The now-superseded NEPA, for example, issued a statement in 1997 accusing foreign investors of 'relocating their polluting industries to China to take advantage of lax law enforcement'.

Chemical plants, dye factories, tanneries, and other polluting projects had been transferred to China because 'they failed to meet environmental standards of their place of origin', it said, specifically naming investors from Hong Kong, Taiwan, and South Korea. NEPA also accused China's own central government agencies of luring industrial polluters to relocate without regard to the environmental impact.

In earlier research, Murray encountered many complaints from Chinese officials of such factory dumping by the Japanese. They were said to be taking the opportunity offered by the opening up of the mainland to foreign investment in the 1980s to get rid of their labour-intensive, low-tech factories whose operations had become virtually untenable due to rising awareness of the environmental damage their were causing. The Japanese, it was claimed, were also giving China mainly outdated, low-grade technology 10 to 20 years old allegedly to hold back the mainland from becoming a serious industrial competitor on world markets (Murray, 1993, pp. 151–2).

If these allegations did once contain a grain of truth, they are long buried by the behaviour of Japanese companies from the mid-1990s onwards in putting more and more investment into operations at the cutting edge of new technology, especially electronics. Nevertheless, some foreign operations have run into problems.

The most high profile case was that of Eastman Kodak, forced to close one of its film plants in central China employing 520 people in July 1999 when new environmental laws to protect a nearby lake made the factory unprofitable. The closure of the medical X-ray film factory in Wuxi, Jiangsu province, was agreed between the US company, its local partner and government authorities, and was not the result of government pressure, said Jessica Chan, a Kodak spokesperson. Kodak bought into the former state-owned plant and formed Kodak (Wuxi) with a local partner in March 1998 year as part of its plan to invest US$1 billion in the mainland. The government, however, tightened laws on waste water disposal to preserve Lake Tai, which abuts the factory. The new restrictions made it 'impossible to operate the plant to be as competitive as other Kodak plants around the world', a company statement said. Kodak had added new water recycling facility and new water pipes, and 'every component of the wastewater treatment plant had been replaced or upgraded. But the effort proved insufficient' (Associated Press 28 July 1999).

5 Urban demographic and consumerist pressures

For years, authorities in the western metropolis of Chongqing had difficulty making up their minds what tree should decorate city streets. Locust trees were replaced with eucalyptus, which, in turn, made way for parasols and then Banyan trees. This, however, had nothing to do with one species being prettier than another, or of city authorities with more money than sense. Rather, they were searching for a species that would not die from the constant battering of acid rain.

'More acid rain is dumped on Chongqing than any city in the world, and there is no hiding from it', admitted Xu Yu, vice-director of the Chongqing Institute of Environmental Science and Monitoring (author interview), blaming the problem on highly sulphurous coal. As has been discussed in Chapter 3, China is the world's largest producer and consumer of coal. About 75 per cent of Chongqing's power in generated by coal, which produces an estimated 800,000 tonnes of sulphur dioxide each year. The pollution is taking its health toll. The incidence of lung cancer in Chongqing is 1.7 times higher than in neighbouring cities, and many Chongqing residents suffer respiratory problems, including asthma, according to Xu. One year, all examinees from Chongqing flunked tests to join the air force because of nasal problems (*Reuters*, 2 April 1997).

Chongqing's population is China's biggest, having doubled to more than 30 million following the city's elevation in political status to a municipality under the central government in 1998, and this adds to its environmental woes. 'Pollution has worsened along with economic reforms', agreed Mayor Pu Haiqing (author interview), citing the fact that the city is one of China's top ten steel makers, as well as of the country's top four automobile and motorcycle manufacturers. There are almost 10,000 factories employing about 1.2 million people, but much of the equipment is old.

This is arguably China's dirtiest city, losing about three per cent of its gross domestic product each year to pollution and ecological damage (Reuters op. cit.). Thanks to air pollution, Chongqing's shoe-shiners do brisk business. Some jobless people carry a can of paint and a roller as they roam the streets looking for business. Paint peels off buildings three times faster than in Beijing due to acid rain, while metal is eaten faster than any other city in the world (see Murray, 1993, p. 76). Box 5.1, drawing on a World Bank study, describes some of the problems Chongqing now faces from rapid urbanisation and industrialisation, especially in the area of water quality.

Chongqing Municipality, originally part of Sichuan Province, was given independent status similar to the municipalities of Beijing, Shanghai and Tianjin. It is now the world's largest single metropolitan area, with nearly 35 million residents, including 6 million urban residents. Chongqing City proper, located at the confluence of the Yangtze and Jialing River, is home to 2.5 million people, followed by Wanxian with 300,000 and Fuling with 200,000. Chongqing has become the largest economic centre in southwest China, growing at approximately 10 per cent a year between 1980 and 1994. The central government's policy between 1987 and 1992 of providing incentives for growth in coastal areas left inland cities behind, and Chongqing's per capita income lagged 30–40 per cent behind other major cities. Now, it is beginning to catch up. Its location at the confluence of the Yangtze and the Jialing Rivers has made it a commercial hub for shipping; it also has a diverse industrial base that includes coal, iron, petroleum refining, metallurgical, chemical and petrochemical operations.

The development of light industry and manufacturing (autos, motorcycles, electronics, food processing, and cosmetics) indicates an attempt to move toward higher end production. Chongqing now forms part of an emerging urban industrial corridor stretching north-west to Chengdu City in Sichuan Province, and is increasingly affected by the Three Gorges Dam construction and related resettlement and infrastructure investments to the north-east. Development in this corridor attracts migrant workers from the surrounding villages, who typically take low-income jobs and do not have legal urban registrations. Once in the cities, these workers have only limited access to public services. As income opportunities in agriculture shrink and urban–rural income disparities grow, the size of this 'floating population' (already estimated at 20 per cent of Chongqing's registered population) will increase substantially, exacerbated by lessening restrictions on inward-migrations. As the number of legal and illegal urban residents grows, urban areas will face increasing pressure to provide affordable and accessible municipal services. Chongqing's economic growth, in terms of population, industry, and agricultural production, has come at the cost of increased pollution of air, water and land resources.

Chongqing is not an isolated case. 'Air pollution in some Chinese cities is among the highest ever recorded, averaging more than ten times the standard proposed by the World Health Organisation. China's six largest cities – Beijing, Shenyang, Chongqing, Shanghai, Xian, and Guangzhou – rank among the most polluted in the world' (World Resources Institute, 1999). 'The residents of many of China's largest cities are living under long-term, harmful air quality conditions', Zhao Weijun, deputy director of the air pollution department of the former National Environmental Protection

Agency, admitted in 1997 (*China Environment News*, 21 January 1997, p. 1). Ambient concentrations of total suspended particulates (TSP) and sulphur dioxide are among the world's highest. In 1995, more than half of the 88 cities monitored for sulphur dioxide were above the WHO guideline, while all but two of the 87 cities monitored for TSP also far exceed the maximum allowable level. Some cities, such as Taiyuan and Lanzhou, had sulphur dioxide levels almost ten times too high (NEPA, 1996, p. 5, 15).

In January 1999, the Washington-based World Resources Institute (WRI) released the results of a study, funded primarily by the World Health Organisation and the US Environmental Protection Agency, aimed at establishing an 'Environmental Health Indicator' by looking at cities in selected countries considered to be posing the greatest risk to the health of children, who are regarded as the most vulnerable to air pollution. The WRI, in its work with the Chinese Government to develop solutions to continuing environmental degradation, used three measurements of air pollution – total suspended particulates, sulphur dioxide and nitrogen dioxide – and found the worst city in China was Lanzhou, capital of the north-western province of Gansu, a region with a large petrochemical industry and oil refineries. 'As a consequence of the emissions from these industries, the city rarely experiences a clear day', it reported. The WHO guidelines state that the maximum permissible amount of particulates is 90 micrograms per cubic litre of air. According to the report, Lanzhou has more than 700. The next worst city, Jilin, in the north-eastern province of Jilin, has close to that, while Taiyuan, in Shanxi province, has nearly 600 (WRI, 1999).

In the 1996 government White Paper on environmental protection, it was admitted that the nation faced an 'uphill struggle in trying to develop its economy without crippling the environment. Pressure on resources, which were already in rather short supply, and on the fragile environment, has become greater and greater'. The report said enormous damage has been caused by pollution, brought about by the growth of the past decade and people's negligence, in which the key factor was rapid urbanisation (State Council, 1996).

The overall problem of air pollution will be dealt with in more detail in Chapter 10. Here, it is brought in merely to set the scene for a discussion of the pressures on the environment exerted by growing urbanisation within the context of a national population that has more than doubled since the founding of the PRC, urban populations swollen by millions of rural migrants seeking a better life, and the growth of consumerism as an increasing number of Chinese have become relatively affluent through the economic reforms introduced since 1979.

Growing urbanisation

Under the old centrally planned economy, the problem of dirty air and water largely came from the belching chimneys of inefficient factories struggling

Box 5.1 Chongqing – urbanisation equals water problems

Within China, Chongqing ranks first among 23 large cities in sulphur
dioxide and eighth for levels of suspended air particles; Chongqing
City is the largest source of organic water pollution in the Yangtze
River Basin upstream of the Three Gorges Dam. Preliminary river
modelling indicates that pollution has severely deteriorated the water
quality in the Jialing and Wu rivers, while so far the Yangtze River has
been able to absorb urban, agricultural and industrial pollution loads
while maintaining acceptable water quality, thanks to its strong flow
and high velocity. Health impacts due to the poor water quality in the
Jialing and Wu, however, have been substantial: intestinal infectious
diseases such as hepatitis A and dysentery have incidence rates some
50 per cent higher than national average, and *E.coli* bacteria are rampant
in some water sources, as high as 15,000 *E.coli*-l in some parts of
Chongqing. The completion of the Three Gorges Dam, some 1000 km
downstream (see Chapter 6), will reduce the velocity of the Yangtze
River, increase its water depth, and alter the flow regime. From a river
basin perspective, Chongqing needs to be concerned with maintaining
a water quality that will avoid eutrophication of the future reservoir.
From a local perspective, Chongqing must ensure that deteriorating
surface water quality, in particular near water intakes, does not
endanger its supply of safe water. In addition, the municipality must
adjust its urban environmental infrastructure to the heightened water
level: for example, two of Wanxian's water supply plants will be
submerged, parts of the sewer system and waste water treatment plants
will be flooded, and open dumpsites located near the river will be
washed out. Wastewater collection and treatment is very limited,
covering only about 13 per cent of the waste stream. Expansion of
water and wastewater management capacity is hampered by the weak
financial situation of the service providers, which are typically water
supply companies and government departments for drainage and solid
waste management.

  These entities find themselves in a transition phase where the
traditional government subsidies have been reduced but the tariffs
for municipal services do not yet cover the full cost of operation
and investment. The municipality now encourages the creation of
financially autonomous municipal companies, but such companies
tend to lose money due to weak financial and operational management,
and relatively high per capita water demand, which in part is the result
of low tariffs which do not induce efficient water use (World Bank,
1999). Some relief is in sight for the sewage contamination of the
Yangtze and Jialing rivers (an estimated 700 million tons of untreated

waste water discharges per year in the late 1990s), however, through
the construction of a 92-kilometre-long underground pipeline network
to collect household wastewater, due for completion in 2004 at a cost
of 2.7 billion yuan. Two treatment plants with daily processing
capacities of 800,000 tons and 400,000 tons will receive sewage from a
72-kilometre-long main pipeline and return clean water to the rivers
(*Xinhua*, 4 February 2000).

to meet the production requirements of the annual 'plan'. The switch to a
market economy, sadly, has brought little relief. In fact, it has encouraged the
growth of more sources of industrial pollution as local governments and
individuals see a chance to get rich.

Although almost 70 per cent of China's 1.3 billion population still lives in
the countryside, the long-accepted view of the mainland as a predominantly
rural, agricultural society needs revision. All the evidence points to China
becoming an urbanised society in a very short space of time, with massive
implications for social stability, industrial structure, economic development,
the environment, and future government policies on a wide range of issues
(see Cook and Murray, 2000, Chapter 7 for a full discussion).

For many years after the founding of the PRC, urbanisation was kept in
check via Maoist policies (see also Chapter 3). Cities were viewed as potential
centres for the evils of capitalism, including prostitution, gangsterism,
poverty and exploitation. Mao's revolution was rural-based, and attention
was therefore focused more closely on peasant concerns rather than those
of urban dwellers. However, the desire for industrial growth helped offset
this bias to an extent, and surplus was extracted from agriculture to aid
industrialisation of the cities. It is debatable whether there is a 'natural' level
of urbanisation, but most experts would agree that in the Maoist era, China
was under-urbanised in comparison to other Third World countries, and
many thought that this was no bad thing given the low quality of life in Third
World cities at that time.

However, although urban dwellers were generally adequately fed, clothed
and sheltered, urban life under Mao was nonetheless austere and Spartan,
and there was a marked lack of concern for the lighter side of life, for the high
and low culture most associated with the city as a form of human settlement
(Cook and Murray, 2000, p. 149).

This situation changed from Deng's rise to power. Since then, the government has sought to encourage rather than discourage urban growth. The total
urban population of China stood at 19.4% in 1980, soon after Deng gained
power, and had grown to 23.7% in 1985 alone (this percentage change may
seem small, but China's large population numbers must be kept in mind),
rising to 26.4% in 1990, 29.0% in 1995 and 30.4% in 1998. These figures

compare to the 17–18% found in the 1960s and 1970s (State Statistical Bureau 1999). Officially, by year-end 1998, those living in urban areas numbered just under 380 million, but many millions of floating population must be added to this total. Conversely, the nature of China's classification of urban areas means the surrounding agricultural population is included in the urban total. This is shown at its most extreme in the newly designated Chongqing municipality where 80% of the 30 million plus population is rural (24 million), but even in Beijing there are approximately 5 million rural dwellers, and 2.5 million in Guangzhou.

In the big cities especially, the changes are dramatic. For instance, the capital Beijing had an estimated population of only 1.2 million within the confines of the massive city walls in 1949. The new government demolished these walls and the population quickly expanded to 4.1 million, including 1.8 million in the city proper (Dong Liming, 1985). By the mid 1980s these figures were nearly 10 million and 6 million respectively, and by the late 1990s Beijing's permanent resident population (having lived in Beijing for more than six months) was nearly 12.5 million (China Statistical Information and Consultancy Centre, 1999).

Beijing's urban fabric has changed in parallel with this increase in population. In the Maoist era the emphasis on heavy industry meant that, even in China's capital city there were 20 steel mills, including the major works at Shijingshan (known as Shougang – see Chapter 8), petrochemical works and other noxious industries, many of these in the downtown area. The great square of Tiananmen was laid out during this period, as were many of the Soviet-style buildings such as the Great Hall of the People to mark the tenth anniversary of the founding of the PRC. Environmentally, the 'excessive' emphasis on heavy industry 'created a shortage of water, electricity and transport capacity, and worsened environmental pollution. Little attention was paid to housing and public facilities' (Zhou Shunwu, 1992, p. 30).

Zhou identifies three main economic changes beginning in 1980:

1 Readjustment of industrial structure, with a greater stress on light industry, a wider range of agricultural activities and township enterprises, and so 'Township industries developed so quickly that their output value took up 49.1 per cent of the gross rural product and 12.3 per cent of the total industrial production' (ibid., p. 31).

2 Construction of public facilities, houses and satellite towns, with satellite cities such as Changping, Tongzhen and Yanshan being established and 52 large-scale housing projects (over 100,000 square metres) and other works being completed.

3 The integration of economic development, science and technology, and education that has not only promoted each of these areas 'but has also made Beijing a centre of technology, information and talents'. The scale of the latter is shown by the fact that by year end 1998, 281,000

were involved in scientific and technological activities, up 2.9% from 1997, and expenditure on research and development was 5.2 billion yuan, up 19% from the previous year (China Statistical Information and Consultancy Centre, 1999, p. 19). Likewise, Beijing's higher education institutions enrolled 62,000 undergraduates and 16,000 graduate students in 1998, and 60,000 graduated in that year (ibid., p. 20).

This urban expansion is exciting and dramatic. In 1998, investment in urban infrastructure reached a then record 32 billion yuan and in the run-up to the celebrations of the fiftieth anniversary of the PRC in 1999, there were over 5,000 construction sites in the capital (China Statistical Information and Consultancy Centre, 1999, op. cit., p. 17; *Business Beijing*, October 1999, p. 18). Meanwhile, in Shanghai, total infrastructure investment was 51.7 billion yuan in 1998, especially in cross-city express-ways plus activities associated with Pudong New Area (ibid., p. 55, and see Cook and Murray, 2000, pp. 152–3, for a summary of Pudong's development).

What of the future? Geographer Kam Wing Chan has projected an urban population of between 35.6 and 39.2 per cent in 2000, rising to 48.2–58.3 per cent in 2010. The numbers involved would be enormous, at up to 838.4 million people in 2010 in a 'worst-case' situation. If these predictions proved accurate, '[w]hat is going to happen in China in the coming two decades is thus truly momentous – the urban percentage will almost double and the size of the urban population added will be about that of the current US urban population' (Kam, 1994). The China Academy of Social Science, which, on various occasions, has projected a 50 per cent level as entirely possible by 2010, supports this thesis.

This growing urbanisation is contributing to an even greater population imbalance between the over-crowded eastern coastal regions, where most of the economic growth has been so far, and the under-populated and relatively economically backward regions in the interior. At the beginning of the 1990s, the average population density was 118 people per sq. km. However, this is unevenly distributed, ranging from a high of 360 people per sq. km in eastern coastal areas, 197 people per sq. km in central areas, and only 13 in the west (Fourth National Census, 1 July 1990).

The area to the east of the line stretching from Heihezi of Heilongjiang Province in the far north to Tengchong of Yunnan Province in the far south accounts for 42.9 per cent of the country's total area but 94.4 per cent of total population; the area to the west accounts for 57.1 per cent of the country's total area but only 5.6 per cent of the population. There are even greater density disparities within regions. For example, the density of population in Jiangsu and Shandong province is respectively 653.6 and 551.6 people per sq. km, while in Tibet and Qinghai the figures are respectively 1.8 and 6.2 (ibid.).

Impact of the market economy

The above figures have important economic development implications when it comes to the allocation of resources, especially in such areas as foreign investment, which has tended to be drawn to the heavily populated coastal areas.

Urbanisation has increased quickly in the past two decades, due to the reform and opening-up policy. Cities have become the backbone of the national economy. Reformation of China's cities has been concentrated on the adjustment of governmental control of enterprises, sharpening competitiveness and improving the state economic structure. It helped shape a market-oriented economy based on socialism. Giving state-owned enterprises increased decision-making power, beginning with six firms in Sichuan Province in the 1980s, was a landmark event in the urbanisation process. A series of flexible policies followed. Establishment of four special economic zones and 14 coastal cities with preferential policies reflected the nation's economic development path. Reforms were broadened when the state decided to loosen control of state-owned enterprises, preparing them for the emerging market-oriented economy. The prosperity of Shenzhen, a coastal city in Guangdong Province that spearheaded the reform, proved the pragmatism of the leadership, then headed by Deng Xiaoping, moved China towards a miraculous economic take-off. The urbanisation process was enhanced when constructive reforms were implemented in virtually every aspect of the urban centres – including distribution systems, finance, housing, welfare, and the social security system. The number of cities grew from 193 to 668 between 1979 and 1997 (State Statistics Bureau figures).

Most cities grew out of rapidly expanding counties. The number of city inhabitants, another barometer, grew from 12 per cent of China's population to 43.8 per cent between 1978 and 1997. The gross domestic product (GDP) in urban areas, a critical economic index, has grown rapidly, achieving an 18 per cent growth rate annually between 1988 and 1996. The mushrooming tertiary industries, a term referring to sectors other than agriculture and industry, leapt from 28.6 per cent to 37.2 per cent from 1988 to 1996. The GDP in 34 cities exceeded 20 billion yuan (US$2.4 billion), according to the State Statistics Bureau (SSB). Bureau reports predict the urbanisation process, while gaining the state's attention, will grow faster heading into the new century.

By 1994, the SSB recognised 622 'cities' throughout the country, 51 in Guangdong Province alone, followed by 46 in Shandong Province, and 39 in Jiangsu Province. In 1993, a total of 32 had at least one million people, playing a key role in economic growth not just at the local level but also at the regional level and, in the case of cities like Beijing or Shanghai, nationally or even internationally. By 2000, the SSB expected to 724 cities, rising to 1003 by 2010 – although it should be stressed, as will be discussed in more detail shortly, that much depends on how one defines a 'city'.

According to Liu Jinsheng, a division chief of the Ministry of Construction, the number of cities had reached 668 by the end of 1998, and registered towns totalled 18,984. 'The number of small cities has increased dramatically, especially on the county level, with eastern areas experiencing more rapid urbanisation than less developed western regions.' A number of metropolitan areas have formed around major cities, including the Bohai Bay area centred on Beijing, Tianjin, Qingdao, Shenyang and Dalian. Another has emerged in the Yangtze River Delta around Shanghai, Nanjing, Hangzhou and Ningbo, and still another in the Pearl River Delta around Guangzhou and Shenzhen (author interview).

Highlighting the unevenness between east and west, Liu noted that in 1995 the state allocated about 77 billion yuan for city construction projects, but only 10.9 per cent went to the western areas. Per capita urban construction investment was 455 yuan nation-wide, but only 295 yuan in Western areas. The gap between eastern and western areas is mainly a result of different population densities and socio-economic development, he explained. The high population density in the east is the result of the relatively richer agricultural land found there. The economic boom has attracted workers and talent from the inland areas, as will be discussed shortly. But the rich natural resources in the vast western areas, of strategic significance to the future development of the national economy, will inevitably accelerate the process of city construction in the west. The government has already begun conducting experiments in western commercial and trade centres such as Lanzhou in Gansu Province and Chengdu in Sichuan, to boost economic development in the areas surrounding these cities.

At this point, it should be emphasised that the urbanisation of China is not something new. At least one study argues that the process has been under way since at least the later period of the Ming dynasty. 'In fields as diverse as higher education and control of water-conservancy projects, almost all of the important institutions and socio-political functions were increasingly concentrated in the cities' (Elvin and Skinner, 1974, p. 3). Nevertheless, it should also be acknowledged that these urban settlements were still relatively small and somewhat manageable. By the beginning of the twentieth century, while 'at least 34 per cent of the population [were] living in settlements containing at last 2,500 inhabitants, only about six per cent [lived] in conurbation's of 50,000 or more' (ibid.).

The urban expansion has been under way ever since, with the mega-cities like Beijing and Shanghai and Guangzhou continue to spread outwards, swallowing up former satellite towns and villages in a featureless urban sprawl with huge populations of 12 to 14 million that have become increasingly difficult to control. This trend is likely to accelerate (Box 5.2).

Until relatively recent times, there was a sharp difference in style between these cities and those of the under-developed interior. Modern cities on the coast began rapid expansion after the downfall of the Qing Dynasty in 1911. 'Between 1910 and 1920, the Chinese population of Shanghai went from

Box 5.2 More mega-cities

Chinese planners continue to dream of ever larger mega-cities, cities with populations big enough to swallow up a Canada of 30 million people with ease, or squeeze in a South Korea of 46 million. They also envisage a network of cities rising from coastal plains and river deltas to eventually absorb a population greater than the 281 million of modern-day America.

This, the planners feel, is the best way to handle the challenges posed by the need to build housing for an increasing population and jobs for the rising number of rural unemployed.

'It's the most economic way to create infrastructure, housing, jobs and pollution control', according to Andy Xie, chief economist at Morgan Stanley Dean Witter in Hong Kong, who monitors mainland developments in this sector. 'With these large cities, things that weren't possible before are now economically viable'. Massive economies of scale, he argues, will allow China to build state-of-the-art infrastructure it might not otherwise be able to afford. 'Telecommunications are going to reach levels never seen before, there will be super-modern transport systems'. Songsu Choi, principal urban economist with the World Bank, says mega-cities will be environmentally friendly, because scattered rural communities are the biggest polluters.

Zhou Yixing, a geography professor at Peking University, envisages 'metropolitan interlocking areas' that devour cities and towns across a vast area. One would be based on the northern cities of Beijing and Tianjin, each with a population of more than 10 million and lying about 110 km apart. 'Beijing lacks abundant arable land and it lacks a port, but Tianjin has both. Linking with Tianjin will make Beijing stronger'. Shanghai would form the biggest megalopolis in terms of area and population stretching around the mouth of the Yangtze river; the Pearl River Delta region, covering Hong Kong, Shenzhen and Zhuhai would be a natural third megalopolis, while a fourth could emerge from the urban sprawl stretching from Dalian to Shenyang in Liaoning province.

Mu Xueming, deputy general manager of Tianjin's Urban and Land Planning Commission, said the planned mega-city would still be divided into Beijing and Tianjin Special Zones, but with unified urban planning and construction. In another option, it would be a belt-like city group along the existing Beijing–Tianjin expressway and railways. Beijing would be redefined more as a cultural and political centre, and industrial enterprises would be removed from the central part of the capital to Tianjin or the area along the expressway to improve Beijing's environment. (Author interviews)

488,000 to 759,000; large office buildings stood next to miserable workers' districts; there were universities, large newspapers, publishing houses a 'Western style of life'. The walled cities of the interior, such as Suzhou on the lower Yangtze and Chengdu in Sichuan, closed their gates at night in accordance with the old cultural traditions. . . . The city was no more than a maze of one-story grey brick buildings' (Chesneaux *et al.*, 1977, p. 119).

Now, as one travels around China, it seems that most of the old cities cannot wait to bulldoze the old and emulate the modernist touches of the metropolises. This is beginning to cause concern, especially as it is leading to the destruction of much of the architectural heritage and culture that is uniquely Chinese. Addressing a Beijing forum on strategic city planning in 1998, the Minister of Construction Yu Zhengsheng issued a strong warning against unrestrained development. He said China would commit an irreversible mistake if rampant construction was allowed to ruin city land-scapes and their cultural and historical values. An international conference hall would be admirable, but not as desirable, if it was built in an ancient village that should have come under a preservation order. Local authorities were torn between economic pressures to have rapid growth and the long-term development of the cities, with the latter often being ignored. Pressure also came from property developers and city dwellers eager to see tangible results of development, such as skyscrapers and highways, sidelining cultural characteristics (*Xinhua*, 1 May 1998).

Mr Yu's experience as mayor of Yantai, in Shandong province, during the mid-1980s added extra weight to his remarks. He said city planning was a teething problem that should command equal prominence to that of agricultural and water concerns. Mayors lacked co-ordinated vision towards urban growth and the absence of legal boundaries aggravated the problem. The minister said that in the previous 12 months, property investment had risen 10 per cent, while the number of construction workers in China was estimated to exceed 35 million.

Enter the rural migrant

Under communism, the cities have enjoyed a chequered career. During the era of Mao Zedong, more than 70 million city dwellers were sent into the countryside to 'learn from the peasants', particularly during the Cultural Revolution (1966–76). This helped give China the lowest rate of urbanisation in the world, at 17.6 per cent, that prevailed into the 1990s.

But urban populations have rapidly swelled in recent years by an influx of rural migrants (see Box 5.3) taking advantage of a relaxation in the regulations on movement around the country and the loosening of the system of residential permits which kept everyone very much in their place. The migrants have also tended to ignore the country's draconian family planning laws allowing only one child per family, adding further strains to the urban social fabric. The migrant issue is discussed in Box 5.5, while Table 5.1

Table 5.1 Number of rural migrants moving
to cities 1990–2010

Period	Numbers (millions)
1990–5	50
1996–2000	100
2001–10	170

Source: World Resources Institute 1999.

shows the size of the numbers involved, with projections up to 2010. The population issue will be dealt with later in this chapter.

Faced with impossible pressures on existing social services, the main cities made strenuous efforts to expel as many rural migrants as possible. Beijing expelled than 300,000 people between the beginning of 1998 and mid 1999, and Shanghai nearly as many, as well as restricting the number of temporary work and residency permits issued. The officially registered population of Shanghai fell steadily from 1994 onwards, and the city had to close hundreds of primary schools due to lack of pupils (*South China Morning Post*, 19 October 1999).

To divert the migrant flow from the major urban centres, since 1980 peasants have been encouraged to work in rural industries, while the government has sponsored the development of small market towns. Under changes to the urban registration system, as many as 150 million peasants who flocked to work in these county towns were reclassified as 'urban residents'. Most of the new urban residents, however, live in towns with populations of fewer than 200,000, enabling official statistics to claim a rather misleading urbanisation ratio of 32 per cent.

In 1997, the State Council ordered a freeze on all further development of these new towns, however, after satellite photographs showed the mainland was losing 500,000 hectares of cropland a year, more than twice the rate that was thought – although much of the land was left unused.

At the time of writing, there was evidence of another U-turn in policy, as the government appeared ready to abandon long-standing opposition to big cities in a bid to reverse steadily declining economic growth rates. Proposals to step up urbanisation, which emerged from a year-long research project by the China Economic Research Institute, were taken up in late 1999 by the State Development Planning Commission, which helps draw up Five-Year Plans. It suggested that in the Tenth Five-Year plan (2001–05), an additional 85 million people should be moved into cities. The arguments for this policy of greater urbanisation, as well an identification of some of the problems that have emerged so far in carrying it out, are summarised in Boxes 5.3 and 5.4, and will be elaborated in the main text shortly.

Xu Lin, a SPDC divisional director, said the commission had concluded that 'a correct urbanisation strategy is crucial to addressing a host of problems facing China, such as sluggish consumer demand and backward

Box 5.3 More urbanisation: the argument for

Many experts advocate accelerating the pace of urbanisation to stimulate domestic demand and increase the rate of national economic growth. According to Liu Yong, an expert with the Department of Development Strategy and Regional Economy under the State Council's Development Research Centre, a one per cent increase in urbanisation annually usually adds three percentage points to economic growth. Liu cited the low consumption level of rural residents as a primary factor affecting sluggish demand. Stimulating demand within this vast population is loaded with potential.

But despite decades of rapid development, surplus labour and over-production of certain key agricultural products, such as grain and cotton in rural areas, have retarded growth in rural incomes and consumption. 'Accelerating urbanisation could play a decisive role in stimulating domestic demand', he argues, predicting that, in the next 15 years, a 30 per cent reduction in the rural population alone could enhance rural per capita incomes by 30 per cent. The existing slow pace of urbanisation (about 30 per cent), compared with the country's industrialisation rate of about 50 per cent, exacerbated the rural labour surplus restrained market expansion.

'Urbanisation and industrialisation must advance in tandem', Liu insists. 'Urbanisation would also spur the country's inert tertiary sector [accounting for 32 per cent of national GDP in 1999 compared to levels of 60–70 per cent in the developed nations]. More rural residents would live in more highly urbanised cities and towns and their living standards would be improved.' Finally, the country's sustainable development strategy would significantly benefit from urbanisation, as Liu estimates 1.2 million ha of farmland could be saved if 200 million farmers left their farms to reside and work in cities and towns.

As far as the Party is concerned, greater urbanisation will lead to a better use of land and other natural resources, while boosting investment spending and stimulate consumption. If each new urban migrant required spending of 30,000 yuan on housing and urban infrastructure construction, this would mean 2,500 billion yuan in capital spending, it calculated. The new urban residents would buy durable consumer goods worth 400 billion yuan and soak up surplus stocks of such items as TV sets, refrigerators and washing machines. (*People's Daily*, 18 October 1999)

Box 5.4 More urbanisation: the argument against

According to an analysis by Gao Shangquan, Chairman of the China Economic Restructuring Society, the urbanisation process currently faces the following major problems:

- Urbanisation is out of line with industrialisation (as mentioned by Liu, above). The problem is most prominent in the development of small towns. Small towns are 'reservoirs' that prevent the excessive flow of the rural population into large cities. During the 20 years of reform and opening, the number of small towns has multiplied seven fold. Undoubtedly they have contributed to the advance of China's urbanisation process. However, the quality of small towns has not increased in parallel with the growth of their numbers. Owing to the lack of scientific planning, the small towns in many areas are sparsely scattered. Most face shortages of funds and serious environmental pollution. Instead of fully playing the role of absorbing surplus rural labour and enhancing agricultural production efficiency, many small towns have withheld the overall development of primary, secondary and tertiary industries in rural areas, causing the serious waste of land and rural resources, limiting the growth of rural consumption and investment demand, and hampering the quality improvement of the rural population.
- Development of medium-sized cities lags behind. In China, the development of medium-sized cities that link up large and small cities and their surrounding areas has remained slow. In general, one large city should be matched with three to seven medium-sized cities. At present, the ratio is obviously low in Qinghai, Ningxia, Guizhou and other western provinces and autonomous regions.
- There is a lack of good planning management. At present, more than 1,000 counties and over 18,000 towns in China lack a complete planning system. The arbitrary construction and development have led to the man-made decline of many county towns. Meanwhile, many cities with a planning system tend to neglect the law governing urban economic operation while drawing up development plans and lack policies and measures to facilitate the dynamic changes in the local economy. Their planning, often divorced from reality, pursues large-scale and new projects, which results in the waste of large quantities of natural and land resources. (Author interview)

rural economy. The guiding principle of urbanisation is to plan and develop super-large and large cities, expand medium-sized cities, and improve small cities and towns.' To bring China's urbanisation level up to the world average, the country will have to turn more than 300 million rural dwellers into urban residents, which could create tremendous investment and consumption demand. He envisaged an annual increase in urbanisation of one per cent for a minimum 15 years period, which would require shifting 85 million of the rural population into cities within the ensuing five years (*China Daily*, 18 October 1999).

But it is clearly recognised that this expansion should not take place in the largest urban centres but, rather, should come through the construction of new towns scattered around the country, particularly in the under-developed and under-populated interior. Zhu Tiezhen, President of the National Small City Development Committee under the China City Development Research Society, pointed out that new townships have mushroomed since the late 1970s. Zhejiang Province, for instance, had 176 towns in 1978, but by 1998 this figure had grown to 965, with a combined population of 14 million or one-third of the provincial total. However, the government needed to develop an overall strategy for their development and impose sound control over the layout of towns. 'The towns in China are generally too small, and many have a population of only 6,000 to 7,000 each. In one place, there is a cluster of 14 such towns along a 30-km highway' (*Xinhua*, 24 October 1999).

The population numbers involved impose tremendous pressures on the urban infrastructure, especially on mass housing provision. Li Quanfu, the Beijing municipal government official in charge of the programme to reduce the migrant population referred to above, estimated the capital in 1999 had 2.859 million residents from other parts of China without residence certificates allowing them to remain legally, of whom 1.8 million were working (see Box 5.5). 'And the number is still increasing, which exceeds our needs and imposes too great a burden on the city's infrastructure and services,' he added (author interview).

Prior to the celebrations of the PRC's fiftieth anniversary celebrations, Beijing authorities carried out a major drive to get rid of 'eyesores' – beggars and street hawkers, especially those operating food stalls regarded as polluting the air with their charcoal fires. Combined, for a few days at least, with the shut-down of factories, the city's air showed a marked improvement, residents generally agreed. But from 2 October it was noticeable that both the beggars and hawkers were beginning to filter back. This prompted the authorities to launch another crackdown that stopped an estimated 20,000 food stall operators from operating. 'To let Beijing preserve its clean, orderly and exquisite environment is the goal of our long-term struggle', declared Mayor Liu Qi. 'This intensive action is only the beginning. From now on, we will continue making big efforts to carry on our work in a relentless and resolute way so as to create a good environment'. (*South China Morning Post*, 2 November 1999).

Box 5.5 The migrant tide

Although the egalitarian principle was espoused through the Maoist era – everyone being poor but happy – in fact this was not absolute. Life was still better in the cities than in the countryside, which remained mired in centuries of backwardness. And in the 1970s, at least 'one was still better off as a man than a woman, as an urban worker than as a peasant, as a state worker than as a collective worker, as a peasant near a large city than as one in a remote area, or as a peasant with many able-bodied adults in the household than as one with several young children'. (Nolan, 1990, p. 129)

Although the Communist revolution had triumphed largely due to the exertions of the peasantry, in the 1950s the Party leadership tended to lose touch with the masses so that 'the gap between urban and rural areas grew in political, ideological and economic terms. An urban-oriented leadership now adopted the kind of economic strategy that is the bane of most Third World countries. Though some attention was devoted to industrialising medium-sized cities in the interior, considerable emphasis was still laid on building up existing industrial areas, such as the north-east and the old treaty ports' (Brugger, 1977, p. 94). The result was an accelerated drift of rural migrants to the cities. Between 1949 and 1958, the urban population increased from some 58 million to 92 million, while a survey of 15 cities found that from 1953 to 1956 the basic population had increased by 28 per cent, and the dependent population by 70 per cent (Schurmann, 1966, pp. 381–2). Since China adopted the new policy of reform and opening to the outside world, the invisible barrier between urban and rural areas has gradually dissolved, but the managing system separating the urban and rural areas has by no means changed fundamentally. In the cities, people enjoyed rights in employment, housing, health care, welfare, insurance, and education according to their permanent residence registration, little of which was available in the countryside. While the main cities were home to the nation's political, economic development, commercial, and cultural centres, gathering most of the social wealth, they remained a dream beyond the reach of most rural youngsters.

The boom in mass media, however, enabled the rural population to share the same information with their urban siblings, and it finally began to occur to them how unfair the existing society was with regard to dispersion of social wealth. During the period from 1985 to 1994, for example, the gap between the per capita income from wages in the urban and rural areas widened from 1.7:1 to 2.6:1. During the same period, the contrast between the urban and rural areas in consumption levels broadened from 2.3:1 to 3.6:1 (State Statistics Bureau data for various years quoted in Murray, 1998, p. 48). When the control on movement was finally eased, there was an explosion of movement towards the bright city lights. Throughout the 1990s, millions of ex-farmers, sometimes with their families in tow, have been on the move, restlessly travelling from city to city in search of a better life. Estimates have appeared regularly in the 1990s of the floating population (*liudong renkou* or 'rootless' people). Based on surveys and a rural census, the State Statistics Bureau in 1997 decided China had an official 'floating' rural population of 80 million, 34 million of whom had worked in cities for more than half a year. But this is almost certainly a drastic under-estimation, and 120 to 130 million is a more likely figure, with the biggest cities attracting up to 3–4 million each (see Chapter 4 of Cook and Murray, 2001 and Chapter 3 of Murray, 1998, for a more detailed discussion of this issue).

A worst-case scenario (see Chapter 11 for a more detailed examination of various future scenarios) has been offered by the California Institute of Technology, in which environmental degradation played the key role. It envisages massive urban migrations, with 50 per cent of the population in urban areas by 2015, rising to 60 per cent by 2025. Entitled 'The Revolt of the Urban Masses', it said in part:

> The migrants came to escape the hard, limited lifestyle of rural China – a lifestyle made harder by growing pollution and environmental degradation – and in search of higher wages and greater opportunities, but they did not all find them. The urban migrants filled huge shantytowns which soon became dangerous, crime-ridden settlements often beyond the control of authorities. Cities, their budgets overwhelmed by the rate of urban growth, could not provide housing, water and energy supplies, or sewage services to the shanty towns.

The newcomers were tolerated as long as the urban economies could expand rapidly. However, once a slowdown occurred but inward migration continued unabated, leading to growing urban unemployment, unrest began to occur. 'The unrest was exacerbated by increasingly dreadful environmental conditions – polluted water, growing toxic emissions from China's massive industrial base, and rising use of coal. During the winter inversions, the sun often could not be seen for weeks on end and virtually everyone wore filter masks.'

Riots in 2025 began in greater Tianjin, when residents of the urban shanty towns marched into the city proper, demanding food, cleaner air and water, and jobs, but quickly spread to other parts of the country, toppling some

Table 5.2 Population growth 1953–1990

Year	National total	Urban	Rural
1953	58,260	7,726	50,534
1964	69,122	9,455	59,667
1980	100,391	20,631	79,760
1990	113,051	29,614	83,437

(Unit: 10,000 people)

Source: National Census.

city and regional governments, before eventually being put down by a combination of force, in the shape of army mobilisation, and promises of wide-ranging reform (Caltech, 1999).

While this scenario could be viewed as somewhat far-fetched, there is undoubtedly the potential for instability. At present, the concerns are still somewhat unfocused. But there is no doubt that within the major cities, there is the worry that the need for high-rise buildings to house growing populations is leading to the rapid destruction of attractive older features. The atmospheric and ancient alleyways of Beijing known as 'hutongs', lined with traditional courtyard homes behind high walls, notwithstanding a few successful attempts at redevelopment in situ, are being overwhelmed by the inexorable growth of mid-rise or high-rise blocks – or even worse, the construction of new traffic-congested roads.

In fact, traffic jams have so tied up the road systems of major cities that at least 15 of them, each with a population of over a million, have either been or are about to be allowed to build subway systems, following a government moratorium of several years due to the high cost experienced in building existing lines in Beijing, Guangzhou and Shanghai. The result is that more of the government budget needs to be poured into the big cities to make them viable places in which to live, making it more difficult for the government to achieve its target of balanced national development. The have-nots in the underdeveloped areas, meanwhile, are then tempted even more to head for the 'golden' cities in a self-perpetuating cycle of ever-greater urbanisation.

Population pressures

China's environmental problems have undoubtedly been made worse by population pressures, and it has become a cliché to recite that 'almost a quarter of the world's population occupies only seven per cent of its arable land'.

The total population would have reached a record 1.5 or even 1.6 billion at the turn of the century at the growth rate prevailing in the 1970s if family planning had not become one of its most significant fundamental policies. The population increased from 500 million in 1949 to 700 million in 1964. In

Table 5.3 Rise and fall of population growth 1950–1998

Period	Population growth (m)	Average annual growth (m)	Annual growth rate per 000
1st Baby Boom (1950–57)	541.7–646.5	13.11	22
Low Ebb (1958–61)	646.5–658.9	3.02	5
2nd Baby Boom (1962–73)	658.9–892.1	19.46	26
Decline (1974–98)	892.1–1,248	14.24	15

Source: PRC Fiftieth Anniversary State Report, 1999.

the mid 1960s, an average of 100 million people were born every seven and a half years. During the following decade, the same population increase needed only five years. Table 5.2 shows the trends drawn from the four national censuses conducted between 1953 and 1990.

Fearing the population would soon outstrip resources and hamper economic growth, the government embarked on a family planning policy featuring strict birth-control known as 'one couple, one child' in the late 1970s. As a result, the birth rate dropped from 30 per thousand during the 1949–70 period to around 20 per thousand in the late 1980s, and further to 16.5 in 1997, despite a peak in the number of women entering the child-bearing age from the late 1980s onwards. Table 5.3 shows the fluctuating population growths over the fifty-year history of the PRC. The decline in the period 1958–61 can be attributed to the famine in that period, which we discussed in Chapter 3.

The government's goals are to maintain tight control so that the population can be held down to 1.4 billion in 2010, finally peaking in the middle of the twenty-first century after which it will begin to fall, According to some mainland experts, such as Professor Lin Fuda, of the People's University, if this can be achieved, 'China might still be able to be self-sufficient in grain and food supply by then'. (*China Daily*, 29 September 1999).

Bringing together the twin issues of a rising population and shrinking arable land – partly from the twin advances of desert and new town construction – in an environmental context, Lester Brown, Director of the Washington-based Worldwatch Institute in Washington, in 1995, first raised the question of 'who will feed China in 2030'. In a number of his writings since then (e.g. *South China Morning Post*, 7 October 1995), he has created a somewhat grim scenario of the Chinese people eating their way through the world's diminishing food supplies as their own ability to produce enough food is eroded by the growing demands of industrialisation.

Murray has already dealt with this in great detail (Murray, 1998, Chapter 5), so Brown's concerns can be treated here in summary fashion. The industrialisation that is raising incomes is simultaneously undermining food production with its claims on cropland, Brown argues. In the southern coastal provinces, where industrialisation is most rapid, land that was until recently producing two or three crops of rice a year is now occupied by industrial parks. This loss of some of China's most productive land has reduced the mainland's rice harvest by seven per cent since 1990. Government plans to find jobs for 100 million unemployed rural labourers require roughly one million factories (assuming each hires 100 workers, which is a good average for township enterprises). Building a million factories and the associated warehouses and access roads will take a vast area of land, much of it cropland. More land will be lost through the country's commitment to developing its vehicle manufacturing industry as a powerful engine of economic growth, leading to the need for far more roads, especially expressways, to house the increased factory output.

'Water may be even more scarce than land', says Brown:

> With the demand for water increasing six-fold since 1949, the northern half of the country has become a water-deficit region. The water table under Beijing has fallen from 4.5 metres below the surface to 45 metres below since 1950. Eventually, these aquifers will be depleted. At that point, pumping will necessarily reduce the rate of aquifer recharge, reducing the irrigated area accordingly. As well as these coming cutbacks in irrigation, farmers are losing water to the cities, some 300 of which face acute water shortages. In early 1994, farmers in the agricultural regions surrounding Beijing were banned from water reservoirs. All the water is now needed for residential and industrial uses in the city, thus forcing farmers to return to less productive rain-fed farming.

He concludes his argument like this:

> With its grain imports climbing, China's rising grain prices are now becoming the world's rising grain prices. As the slack goes out of the world food economy, China's land scarcity will become everyone's land scarcity. As irrigation water losses force it to import more grain, its water scarcity will become the world's water scarcity.

The water issue will be dealt with in Chapter 7. Here, we can deal briefly with some of the Chinese counter-arguments. They are led by Nong Roan, a member of the Soft Science Committee of the Ministry of Agriculture, who dismisses the dark scenario as lacking any scientific basis. He argues that the conclusions of Western analysts like Brown are based on the experiences of Japan and the Republic of Korea, who both imported large amounts of grain due to the reduction of domestic output during their industrialisation. 'It is impossible for China to rely on others to feed its huge population. The Chinese government has long followed a basic policy for being basically self-sufficient in grain, and it will continue to make unremitting efforts in this regard.' Loss of current arable land, Nong insists, can be offset by better methods of crop production and the potential for utilising land presently considered as marginal (author interview).

Jian Song, Chairman of the State Science and Technology Commission, is equally reassuring. For a population that may eventually reach 1.6 billion, he anticipates a need to produce 640 million tons of grain annually to meet basic needs, based on per capita annual consumption of 400 kg. And this, he believes, can be attained by a mere one per cent annual growth in grain output each year (China claims 3.1 per cent annual growth in the past half century). Use of improved plant varieties, and genetic engineering, will increase harvest yields, while new agricultural techniques can open up current marginal land for harvest purposes (author interview). These upbeat assessments nowhere consider the environment to be a problem. This may be tempting fate, as we will consider in due course.

Consumerism and waste disposal worries

No matter how successful the government may be in this effort, the simple fact remains that China has a vast population, and one that is becoming increasingly consumer oriented. Western businesspeople from the nineteenth century onwards have dreamt of the untold riches to be gained from supplying such a vast market. But, to give one illustration, if Coca Cola is able to supply each Chinese citizen with one bottle or can of its famous drink, that amounts today to 1.3 billion bottles or cans that will somehow have to be disposed. Multiply this, then, over a whole range of products that form the daily consumption of the average person, and one is left with a waste disposal problem of frightening proportions.

Consumerism has certainly arrived in China and is already causing headaches in waste disposal, even though it is still at an embryonic stage. According to the Gallup Organisation, the self-reported average family income overall was 10,400 yuan a year in 1998, a 74.5 per cent increase over the 5960 yuan reported in 1994. Urban families said they made an average of 14,000 yuan per year and had disposable income of about 2500 yuan. Families in the southern economic hothouse of Guangzhou, however, reported incomes twice as high as the urban average (Gallup, 1998). Most city residents, therefore, have sufficient money to indulge in some form of extravagant consumer spending, and a growing number of peasants are in the same position.

Most Chinese households now own major consumer durables. Gallup found nine-tenths of Chinese households already possessed televisions and rural TV ownership was getting close to the national average. Ownership of colour televisions, for the first time, surpassed that of black and white televisions in 1998. Most families now own a radio, tape recorder, stove or cooking range, bicycles, electric fans, a steam iron and washing machine. One in every four Chinese households owns a refrigerator and telephone, but only a small proportion of Chinese households own high-end goods such as pagers (15 per cent), motorcycles (14 per cent), video cassette recorders (12 per cent), air-conditioners (6 per cent) and mobile phones (4 per cent).

Among the largest discrepancies between short-term and long-term purchase intentions were for cars, computers and mobile phones. Only one per cent hoped to purchase a car in the short-term, but 11 per cent held this out as a longer-term objective. Four per cent planned to purchase a computer in the near future, but 16 per cent hoped to in the long-term. Although only three per cent said they intended to purchase a mobile phone within the next two years, five times as many cited this as a product they would eventually own (Gallup, op. cit.).

Until the 1990s, private car ownership in the PRC was seen as indicating bourgeois capitalist tendencies; later, it was considered an extravagance, rather than a necessity. But, since April 1993, the State Planning Commission and other interested bodies have been working on a strategy for the development of a car industry eventually to be dominated by private ownership. Although at the time of writing ownership remains dominated by the public sector, private ownership is surging with the emergence of a genuine urban middle class, along with some rich farmers.

The government projected a potential market of up to 4.7 million buyers by the turn of the century, rising to 16.5 million by 2005 and 40.5 million by 2010. In more realistic terms, perhaps, comparing the experiences of other countries, total demand for new cars may reach 4.4 million by 2010, by which time there will be four vehicles for every 100 families in the country. Most of them will have been produced through collaboration with foreign manufacturers.

An indication of government thinking can be gleaned from the following:

> The growth of the motor industry will give impetus to the overall growth of China's economy. In the history of economic development, a common occurrence is that, once average income reaches a certain level, individual automobile purchasing becomes a driving force of economic growth. Having more people buy cars can be done by shifting consumption from food and clothing to housing and transportation. Less than 1 per cent of urban households have cars, which means there is still a huge market for car sales. With the auto industry having the potential to command a greater presence in the nation's economy, [it] should be promoted as an engine of economic growth, not only increasing tax revenues but also providing more employment opportunities.
>
> (*China Economic Times*, translation
> in *China Daily*, 30 October 1999)

But given the high level of air pollution in major cities as mentioned at the beginning of this chapter, largely blamed on exhaust gases, at a time when car ownership is still so low, we are entitled to ask: what will it be like when ownership levels reach those of the West? Of course, one can argue that the technology will be available to mitigate the negative influence of exhaust gases, but it still raises questions as to how all these vehicles are to be fitted into the available road space without total redesign of urban areas and unacceptable levels of encroachment on existing agricultural land.

'In China, everybody wants a car now. It has become a consumer item like having a television or a fridge', says Hu Yan, deputy editor of *Auto Fan Magazine,* citing a survey in Beijing which found that 90 per cent of people wanted to buy a car and 59 per cent believed they would be able to do so before 2010. 'People have discovered the long weekend. They want to go out for a drive and get away from congested cities' (author interview).

But, at the same time, he feels that 'China is not ready for a family car and may never be. It's all a dream.' A mass market would be a nightmare, he suggests, given the amount of pollution generated and the inability of the projected road network to cope with demand and urban congestion, so that, sooner or later, the government will have to take tough measures to curb car ownership even before the mass market has begun to emerge.

The *China Economic Times* article already mentioned sought to deal with this issue in its conclusion:

> Some people worry that a push to develop China's auto industry will mean more traffic congestion. The solution to traffic woes lies in developing a fast and convenient transportation system – not by limiting auto sales. In recent years, stringent emission control standards have been adopted in some major cities to limit auto pollution. However, as more Chinese purchase cars, there will be a more mobile population. In view of this, the interests of environment protection and car sales should be balanced. Simply introducing all the emission standards used in developed countries will create a hurdle to the sound development of the auto industry. Before the implementation of any emission regulations, both consumer purchasing power and the overall interests of the auto industry should be taken into consideration.
>
> (op. cit.)

It would seem somewhat hard on the Chinese populace to be nagged about their profligate consumption, just when, perhaps for the first time ever, they actually have money to spend and quality goods available to spend it on. Most Chinese who are adults will have some memory of privation, and of fearing criticism for indulging in even the slightest display of consumerism. Nevertheless, the ecological crisis is already creating new constraints.

The official media regularly take citizens to task, urging them to accept more responsibility for smog-choked air and polluted water. For example:

> If every individual Chinese willingly reduced the waste of water and electricity, used recyclable products instead of shopping bags and food cartons, sorted their recyclable waste and refused to eat wild animals,

the 1.2 billion Chinese would surely become the world's most mighty environmental protection army.

(China Daily, 22 March 1999)

Faced with a sharp rise in the amount of urban garbage produced, waste treatment and utilisation have become hot topics. China has traditionally buried solid waste and made compound fertilisers from organic waste. Some areas have introduced incineration methods and used the resulting heat to provide power, according to Zhang Jinfeng, an official and researcher in the Environment and Sanitation Engineering Technology Centre of the Ministry of Construction. Now, waste treatment in urban areas, which previously consisted of street-cleaning and shipping garbage to landfills in suburban areas, is seeing a revival of the recycling concept. Lack of co-ordination has reduced the efficiency of trash utilisation, however, and, says Zhang, China's current regulations and disparate waste disposal and recycling methods are not ready to face the coming revolution.

> Before the 1980s, people sent waste paper, metal products, glass bottles, toothpaste tubes and plastics to waste recycling centres supported by a large workforce. The recent increase in the amount and types of waste are creating a heavier environmental burden and are making collection more difficult. Trash collection remains a fairly backward operation, dominated in Beijing, for example, by individuals who tool around on three-wheeled cycles outfitted with a garbage receptacle. Itinerant workers from outlying provinces often handle the difficult and dirty work. Waste that cannot be sold for cash is generally disregarded and ends up in landfills.

Although the State has increased investment in urban garbage treatment for the next few years, Zhang argued that policies needed to be improved to ensure effective management For instance, local governments should give subsidies to cities that collect and treat inorganic and organic waste in accordance with a standardised classification system. Government authorities could also reduce the amount of non-recyclable garbage by imposing taxes and other disincentives on the use of certain disposable products, especially those that are difficult to recycle. These tax revenues could be used to fund garbage treatment programmes. Local governments could also give tax breaks to encourage the development of the garbage collection and recycling industries.

A strict classification system in countries such as Switzerland and the Netherlands has been credited with high rates of recycling. Zhang disagrees with the common view that China lacks the funds to establish a classified collection system and make recycling a success. First, compared with building large garbage treatment plants, it is more efficient and cheaper to use a classified collection method. Further, successful pilot projects in Beijing,

Shanghai and Shenzhen indicate that a classified collection method is workable. As part of an enhanced environmental protection policy, special attention should be paid to the treatment of toxic and dangerous wastes, such as batteries, which should in turn reduce heavy metal pollution. Successful programmes in other countries indicate that proper garbage management and treatment requires public support. Publicity activities could include environmental education courses in schools and public announcements on billboards, TV and radio (author interview).

According to one senior lawmaker, at least half of the mainland's cities are besieged by garbage, with Beijing surrounded by 4,700 rubbish dumps (see Box 5.6). Professor Qu Geping, head of the environmental protection committee of the National People's Congress (NPC), told a Hong Kong conference organised by the United Nations Development Programme that the mainland desperately needed to speed up the recycling of urban waste.

Box 5.6 Coping with Beijing's waste

In 1950, the new government decided to remove the capital's rubbish tips that had been steadily accumulating since the nineteenth century. A total of 7,000 labourers, 800 trucks and 30,000 animal-drawn carts were mobilised and in 61 days an estimated 600,000 tons of rubbish were removed from the existing inner city area. This was, however, merely to new sites in what were then the rural outer suburbs. As Beijing grew, the rubbish piles were continually resited further and further out into the new outer suburbs – all the time growing in size. By the mid-1980s, satellite remote sensors established the presence of 4,700 rubbish dumps, each with a minimum diameter of 50 metres, in a giant ring around the city's outskirts. In the 1950s, Beijing generated two million tons of household rubbish a year. By 1998, it was at least double that, with the much-enlarged city generating an estimated 11,000 tons of waste a day. But only just over two per cent was being processed. At this stage, the municipal administration launched an ambitious waste disposal program to treat 60 per cent of waste by the end of the century.

The first step was a 60-hectare landfill site developed with loans from the World Bank at a location 35 kms from the city proper. This was opened in mid-1994, with a daily handling capacity of 2,000 tons. Special trucks pick up the garbage around the city, spray it to kill germs and compress it into a two metre high layer for delivery to the site. Waterproof sealant has been applied to the site foundations to prevent toxic liquids leeching out to penetrate the underground water supply.

(*continued overleaf*)

Deep wells have been sunk at regular points nearby so that water samples can be taken constantly to check for contamination. With a projected capacity of nine million cubic meters, at present rates this facility will reach saturation point in 2005, whence an artificial hill 40 metres high will be planted with trees and the site turned into a park. But this is still just a more advanced version of the old method of permanent rubbish dumps around the capital, and Beijing environmental experts acknowledge that it is only a stopgap measure. The future, they say, lies in finding ways to either eliminate the waste or convert it into something useful. In March 1995 a small pilot plant opened to convert selected rubbish into organic compound fertiliser through a process of repeated fermentation and magnetic separation. In July 1996, a pioneer furnace burning rubbish was opened using local technology supplied by the Beijing Modern Rubbish Disposal Engineering Co. Using a core sintering technique it aims to create cinders to be used as a light building material. Waste gases are scrubbed to prevent air pollution. The facility has a daily capacity of 500 tons.

Zhang Guolin, Director of the Planning and Infrastructure Protection Agency in the Beijing Municipal Government, says the city invested 1.3 billion yuan in domestic rubbish disposal in the period 1991–5 to render garbage harmless. As a result, the rubbish that met the criterion of 'no air or water pollution threat' rose from two per cent in 1993 to 21.2 per cent in 1995. By 2004, said Zhang, three compost-making factories, four comprehensive recycling centres and two power stations burning rubbish to provide the necessary heat, would be completed to cope with the disposal of 2,100 tons of rubbish a day. (Adapted from Chapter 9 of Cook and Murray, 2001)

Mainland cities produced more than 130 million tonnes of waste in 1999 and this would continue to increase by 10 per cent a year. A typical city dweller produced 1kg of waste a day, while output for residents of affluent cities such as Shenzhen was twice as much. The problem was getting out of hand, partly because local governments were not implementing rules requiring urban residents to sort out rubbish for recycling, he claimed.

'Beijing used to sort urban waste into several categories for recycling in the 1980s. Nowadays, people simply don't bother.' Professor Qu said an army of more than 100,000 scavengers, mostly rural labourers from Zhejiang and Henan provinces, had prevented the problem spinning out of control in Beijing, but 'scavengers only sort out large pieces of rubbish such as scrap metal and sell them. They don't care about small pieces of garbage. We should have organised them better so they could contribute more in handling urban waste' (*South China Morning Post*, 20 December 1999).

Garbage disposal will be considered again in Chapter 10, when we come to discuss the various initiatives being taken at the national and local government levels to clean up the environment. But in order to reach that point, we now need to turn out attention to another key issue – namely, the vagaries of water supply across the vast territory of the mainland.

6   The Sanxia dam

Dr Sun Yat-Sen first suggested the hugely controversial Sanxia (Three Gorges) Dam Project in the early years of the twentieth century, perhaps as early as 1918 (see Box 6.2). Many years later, as we have noted elsewhere, 'Mao gave the project his wholehearted support and the 1950s witnessed immense research into the possibilities of holding back "Wushan's clouds and rain"'. The financial cost was deemed too high, however, and it was not until the 1980s that the plan was resurrected' (Cook and Murray, 2001, p. 208).

Eventually, following considerable political infighting and dissent, this, the largest dam and hydro-electric power project yet conceived, was officially approved in 1992. The first phase was completed in November 1997, with the river being dammed. It is now set for completion in 2009. The dam itself is questionable, and, as we shall see it has high opportunity costs. Not only has this mega-project tremendous domestic implications; it is also globally significant in terms of the type of future that people generally wish to see. In analysing the pros and cons we shall organise our discussion into sections on justification, then critique and finally an assessment of the debate.

Justification: responding to threats

In this section we consider the arguments in favour of the Sanxia project, beginning with that perennial threat to China, the danger of floods, and the consequent need for flood prevention. Box 6.1 details some of the major floods in the Yangtze valley in the twentieth century. Although, as we have seen in Chapter 2 and will return to in Chapter 7, it was the Yellow River became known as 'China's Sorrow', in the twentieth century the Yangtze became the more threatening. The figures in Box 6.1 are extraordinary, with thousands of deaths, huge areas inundated and millions of people affected, all on a regular basis. For example, 'The flood of the Chang Jiang [Yangtze] in 1931 was probably the most disastrous ever recorded in the world. In 1991, another disastrous flood caused great havoc in the region' (Zhao Songqiao, 1994, p. 227).

Box 6.1 Examples of severe Yangtze floods, twentieth century

1910–11   Hundreds of thousands of deaths, millions of refugees, and millions of acres of crops flooded in Yangtze and Huai valleys (Spence).

1931   An estimated 14 million refugees in an area the size of New York State inundated (Spence). More than 50 million people directly affected, 88,000 sq. km flooded. 10 million lost their homes (Kolb). Probably the most disastrous flood ever recorded (Zhao).

1954   More than 33,000 people killed (*South China Morning Post*, 11 August 1998). The floodwaters were diverted, flooding the homes of more than a million people in Jianli and Honghu (Li). According to the current vice-minister of water resources, Zhou Wenzhi, it was these floods that were 'the worst this century' (*China Daily*, 28 April 1999).

1991   Total of 230 million people affected, 50,000 injured, and 1.2 million hectares of crops lost across the nation, including the Yangtze and Huai valleys, their tributaries, and rivers in the north-east. Financial loss 82,100 million yuan, equivalent to a quarter of the Chinese budget for 1991 (Edmonds).

1998   Official data shows that by 22 August, 3004 people had lost their lives (some suggest many more), 1320 along the Yangtze alone. 21.2 million hectares of land in 29 provinces, municipalities and autonomous regions were flooded, with 13 million hectares of crops ruined, affecting 223 million people and destroying 4.97 million houses. Direct economic losses reached at least 166.6 billion yuan (Li) or up to 255.1 billion yuan (*China Daily*, 28 April 1999, op. cit.).

1999   At least 3300 people killed and 13 million people evacuated.

In 1931 at least 10 million people lost their homes, there were 14 million refugees, and 50 million people were affected. Sixty years later, the 1991 floods were significant not so much for the loss of life, which was thankfully less, but for the sheer numbers of people affected, not just in the Yangtze valley but elsewhere, and the cost to the PRC. It was this flood that finally tipped the balance in favour of the Sanxia project. The authorities felt that something drastic had to be done, and the official go-ahead was therefore given the next year, at the National People's Congress – although this was more of a rubber-stamp, for work had already started by that time.

The genesis of the Sanxia project is usually traced back to Dr Sun Yat-Sen, in the 1920s, but Yabuki shows that Sun referred to this vision slightly earlier in his 1918 *Strategy for National Construction*. Box 6.2 shows the long drawn out process of debate and discussion, involving overseas experts such as Dr John Savage in his 1944 feasibility study, as well as Chinese experts themselves. In the modern era, the 1980s were a key time, involving a host of government agencies from the central to the local level, CPC interest groups, many research institutes, and many experts from China and abroad, including the Ministries of Finance, Water Resources and Electric Power, Electronics, Communications, Machine Building; governments of all affected provinces, including Shanghai Municipality; Chongqing City and all urban areas potentially flooded or chosen as resettlement sites; the Yangtze Valley Planning Office (12,000 staff in 1985); members of central ruling group, their staffs and key commissions in Beijing; 58 units and factories specialising in relevant research, design and construction; 11 research institutes and universities; 'numerous' consultants and entrepreneurs from USA, Japan and elsewhere (Spence, 1990, pp. 695–6).

It was small wonder, with such an array of interest groups, that the decision-making process was so drawn out, and the decisions as to its overall feasibility, dimensions, precise location, and the resettlement process were continually delayed. Opponents included the economic construction group of the All-China Political Consultative Congress, whose conclusion that the dam would produce more damage than benefit was one 'that delivered a great shock' (Yabuki, 1995, p. 197) and was in turn refuted by the advisory group of the Three Gorges Development Corporation. Tiananmen and its aftermath probably played a part in cutting this Gordian knot, for on the one hand it suppressed dissent generally, and on the other it strengthened the hands of Li Peng and Jiang Zemin, both known supporters of the project. The power of the central ministries increased relative to the intellectuals and others, from the Chinese Academy of Science for example, who opposed the dam, and as Bradbury and Kirkby have shown, the latter could not resist the vested interests of the former (Bradbury and Kirkby, 1995). The 1991 floods gave the final push.

The increase in flood frequency and severity noted above is largely due to the population shifts in China, as millions of people have moved southwards from their original base in north China. These numbers put intense pressure on this new environment; as was shown in Chapter 2, wood was utilised in great quantities for urban construction, fuel and ink, and woodlands were cleared for agricultural expansion and for reasons of increased security. This large-scale deforestation has increased the vulnerability of the Yangtze valley to floods. Soil erosion has increased, so that silt levels have increased markedly; by the mid-1980s the Yangtze had an average annual sediment load of 486 million tons compared to the 1.12 billion tons of the Yellow River, and more than six times that of the third-placed Zhu Jiang (Pearl River) at 83 million tons (Douglas *et al.*, 1994, p. 190).

Box 6.2 The chronology of the Three Gorges Project

1918	Dr. Sun Yat-Sen writes of the dam in his *Strategy for National Construction*.
1932	Geological studies completed by Guomindang government, moving towards site selection.
1944	'Savage Plan' produced by US expert Dr John Savage, after an on-site investigation.
1958	CPC Central Committee meets in Chengdu to assess Three Gorges water conservancy and long-term plan for the Yangtze.
1970s	Gezhouba dam completed downstream from the Three Gorges.
1984	State Council commission feasibility study; report adopted.
1984	November: Premier Li Peng announces foreign involvement would be 'welcomed'.
198	Spring: Strong opposition voiced at a meeting of the All-China Political Consultative Congress.
1985	July: After a 38-day on-site investigation the economic construction group of the Political Consultative Congress (PCC) reports that the dam would 'produce more damage than benefit'.
1985	October: The PCC report is refuted by the Construction Preparation Advisory Group of the Three Gorges Development Corporation.
1986–91	Stalemate.
1991	Severe floods on the Yangtze strengthen the hand of those in favour.
1992	March: Vice Premier Zou Jiahua, director of State Planning Commission, officially proposes construction.
1992	Spring: NPC approves the proposal.
1993:	Construction begins.
1997	November: First phase ends with the damming of river.
2003	Target date for the first electricity to be produced.
2009	Target date for completion.

It is not just floods that are a major concern. Another major threat is to the economy via energy shortages in the Shanghai area, perhaps especially in Pudong. The hydro-electricity produced will have a dramatic effect on energy production for the Yangtze delta, and is another major attraction of the Sanxia project. It has also been suggested that it will facilitate the quintupling of river freight traffic, thus enabling Chongqing's port facilities to grow further and faster, with 10,000-ton vessels being able to sail from Shanghai to

Chongqing. This will ease the strain upon the overstretched railway system and improve development in the interior of China, helping to offset the regional imbalances currently bedevilling China's development path.

However, as Cook and Murray (from which the subsequent details are taken) note, it is 'the sheer scale of the project which is breathtaking' (2001, op. cit. p. 208). This is the biggest dam yet to be built, and possibly the largest ever. It will be 2 km long and 188 m high, creating a lake an incredible 603 km (375 miles) long with a water level of 175 metres (for a technical discussion of the construction see Freer, 2001). Eventually, electricity will be produced (target date 2003) and the project completed by 2009. Proponents of the dam note that hydro-electricity production will be 18 gigawatts p.a., the equivalent of 18 nuclear power stations or the burning of 100 million tonnes of coal, and will thus save China from the environmental pressures of these alternatives. This electricity is set to meet 11 per cent of China's total energy needs.

> Chinese officials argue that without the dam, in a major flood similar to those of the past, 'over 600 million hectares of high-yield farmland would be flooded and the lives of some 5 million residents would be in jeopardy' (Hing Qingyu, Yangtze River Water Conservation Committee, in Pang Bo, 1998). Similarly, without its hydro-electric power, the equivalent burning of coal would produce 370,000 tonnes of NO_x, 2 million tonnes of SO_2 and 10,000 tonnes of CO (Zhoa Bian, 1998). The dam will thus reduce the risk of acid rain, 'help curb the global greenhouse effect and thus contribute to environmental protection' (Wei Tingcheng, State Council Three Gorges Construction Committee, cited in Zhoa Bian, op. cit.). Farming incomes in the area affected were half the national average and so the resettlement process is viewed as 'providing . . . a historical opportunity to move from poverty to wealth' and the migrants 'will be guaranteed a happy and peaceful life'.
>
> (Wu Bian, 1997)

Critique: mega-project dangers

Without doubt, the dangers of floods in the Yangtze valley are immense and have increased markedly in the twentieth century. Clearly, something major has to be done in order to deal with this perennial threat. However, critics of the dam project (and it is indeed hard for us to stop adding the letter 'n' to 'dam' here) are appalled, believing the 'sheer scale' will endanger the Yangtze valley environment and its people. To them, 'This is the most socially and environmentally destructive infrastructure project in the world today' (Lammers, cited in ibid.). Critics are concerned with such threats as:

- Raw sewage or toxic waste pumped into the reservoir
- Pollution from the 657 inundated factories, which will not have been properly cleaned

- Backwash from the reservoir in the event of a flood, threatening the lives of 500,000 people upstream
- Sedimentation and siltation clogging up the reservoir itself, and also leading to Chongqing's harbour becoming unusable
- Siltation and the reduction in river flow endangering the rare white dolphin, giant salamander, and other animals
- An earthquake, triggered by the weight of the dam, leading to massive loss of life in the ensuing flood
- Vulnerability to attack by terrorists or outside powers in the event of war, with huge potential loss of life in the subsequent flood
- Corruption of local officials and builders, endangering safety standards in construction (as has happened elsewhere in China), thus leading to potential cracks in the structure of the dam, again with a huge potential threat to life if it gives way. Concerns over corruption led to foreign quality control experts being called in from May 1999 to report directly to a committee headed by Zhu Rongji (*Associated Press*, 31 August 1999).
- Impact on the 1.2 million (some say 1.9 million) people being resettled, for they are, contrary to the happy picture of prosperity noted above, 'depressed by the economic loss they will suffer and disturbed by the inevitable break-up of the emotional ties they have had with this land' (Wu Ming, 1998)
- The immense cost of the dam, which was 90 billion yuan at one stage but may be as high as 240 billion yuan
- Loss of precious archaeological sites that will be inundated, and of the ten towns to be flooded.

Some of the most trenchant critiques have come from experts with the International Rivers Network based in California, such as Owen Lammers cited above. During the 1998 floods, Philip Williams, president of this network, noted that 'What's happening now is a vivid illustration of the failure of China's flood control policies', and argued that the dam would not reduce the risk of flooding in areas that for the previous few weeks had been drenched (*Agence France Presse*, 8 August 1998). This is due to the problem of sediment deposition behind the proposed dam. This 'will rapidly reduce its capacity to store flood waters', the network said. Chinese critics agreed. Ecologist Liang Congjie from the Friends of Nature group was quoted as saying that 'We will only be able to control one-third of the water' (ibid.). The main tributaries of the Yangtze, the Li, Yuan, Zi Qing and Xiang, would continue to flood into the river in the most vulnerable areas of the basin.

Not only are there such questions about the capacity of the dam to control floods in the long term (see Dai Qing (1998), another Chinese critic who has been widely referred to in the West), there are short-term dangers while the dam is being built. The blocking of the river in November 1997 meant that a parallel 3.6 kilometre diversion channel was built to carry the waters. But, 'In order to protect the coffer dam at the construction site, flood water has to be discharged at maximum capacity', according to Professor Huang Wanli from Beijing's Qinghua University (ibid.). 'This puts tremendous pressure on the lower reaches of the Yangtze'. He blamed increasing deforestation around the river for the floods. The lack of trees meant more and more soil was washed into the river, causing sediment build-up. Residents living along the banks of the Yangtze are forced to build ever higher dykes to contain the seasonal summer floods and compensate for the raised riverbed, contrary to the ancient techniques described in Chapter 2. Opponents of the dam project have charged that the authorities are putting more effort and funds into the scheme and ignoring dyke defences.

As regards pollution, a CNN Earth Matters report in 1999 noted views that:

> Instead of a scenic man-made reservoir, [critics] foresee a stinking effluent lake filled with raw sewage and industrial chemicals backing up for 600 kilometres (372 miles). Silting, they say, will block river drainage outlets in Chongqing: sewage will bubble up through manholes and slosh through the streets of China's largest city.

One billion tons of sewage is the official estimate for entry into the lake, a problem compounded by the restriction on water flow imposed by the dam, one which will impede the natural self-cleansing action of the river, already badly affected by current levels of pollution along its length and breadth. Local officials say that they are alive to this potential problem, but the CNN report carried the view of Ma Shulin, vice-chairman of Chongqing's Planning Commission that although they 'will work to reduce the possibility of eruptions in the waste water system', nonetheless a blow-back effect was 'totally possible'.

Then there is the impact on the people who are to be resettled, people who are being exhorted to move for the good of the nation. Li Peng himself called in October 1999 for the process of resettlement to be speeded up, for 450,000 people had to be resettled by 2003 alone. These people would face undoubted psychological upheaval in moving from what for many would be their ancestral home. But at least the original plans envisaged relocation of most people to neighbouring areas. By February 2000, however, in response to Li Peng's exhortations and to Premier Zhu Rongji's inspection tour in May 1999, in which he declared the original resettlement plans were impossible due to the inefficiency or heavy pollution of the TVEs that were projected to be the new employers (*Agence France Presse*, 19 February 2000), reports were coming in via Xinhua and Agence France Presse that migrants were to be relocated far away from their homes, for instance on Chongming island off Shanghai. A group of 5500 migrants from Yunyang county, Chongqing, were set to arrive within the next two or three years, being resettled among the island's 48 villages, to work on farms or in TVEs, the reports indicated (ibid.). Qi Lin, director of the Yangtze Dam's Relocation and Development Bureau admitted that an extra 42,000 people would be relocated in 11 provinces distant from the dam, bringing the total to 125,000 of those who would be resettled far from home. In all, from 1993–99, 220,000 people had been resettled in the area, and preparations made for movement of another 110,000. The total budget for resettlement 1993–2009 is 40 billion yuan (*Xinhua*, 17 February 2000).

For a comparative perspective, Box 6.3 provides details of effects that have been observed with other major dams elsewhere in the world. Some of these have already been noted above as likely for the Sanxia dam, others too are possible. Controversy over dam construction in another country was perhaps most evident in the case of the Narmada Valley Development Project in Madhya Pradesh in India, which involves a series of dams at a total cost of $5 billion (Vidal, 13 January 1998). The World Bank was to provide much of this funding, but eventually withdrew due to local and international protests about the impact of this notable mega-project.

Assessing the debates: alternative ways forward

The arguments in favour and against are powerful, and powerfully put. In assessing these arguments, we suggest that the sheer scale and impact of this project means that it is not enough merely to adopt a rational 'scientific' analysis, important though that may be. It is also necessary to consider some of the less tangible and more emotional or spiritual aspects of human–environment interaction. In previous chapters we have illustrated the dangers of an over-reliance on 'man conquering nature' approaches, such as the central role of the state in Maoist times, or that of the market in Dengist times, while similar examples were noted in the historical record, such as the construction of the Great Wall, or of China's great cities (see Chapters 2 and 3). Table 6.1 summarises factors in favour of the Sanxia project, while Table 6.2 summarises those against.

Table 6.1 covers six dimensions: economic, political, historical, spatial, social and environmental. The emphasis is on control, power, growth, and the sheer dynamism of humanity (although perhaps we should say 'man' here). This is the *gung-ho* spirit of can-do, of wrestling with and thus overpowering nature. It is a rather Western tradition, much associated with the USA, and also with the USSR, although the examples of the Great Wall, the Grand Canal and other great projects of the past show that China, too, has such a tradition. In this 'model', the state, and within it the Communist Party of China, is all-powerful, and seeks to tame the environment on behalf of, and for the good of, the mass of the people. It fits the PRC's history well; it is hard to imagine its leaders, Mao, Deng or Jiang, notwithstanding their different emphases, quibbling with such a mega-scale approach – even if Mao would not have countenanced a role for 'capitalist roaders' in such a project.

All these leaders would no doubt have regarded those who drew attention to the negative factors as 'counter-revolutionaries', 'feudal remnants' or

---

Box 6.3 Global comparisons with dams elsewhere

Known as 'surface impoundment', it is estimated that worldwide there are 41 artificial lakes above 1000 sq. km in area, including Lake Nasser in Egypt with 5,300 sq. km of area and the massive Volta Dam area in Ghana with 85,000 sq. km (Simmons, 1996, p. 294). The environmental consequences of these include:

- Flooding upstream
- Aggradations of upstream channel
- Delta formation upstream and siltation from inflow
- Chemical changes in the vegetation and soils of the drowned area, affecting water quality and fish species
- Evaporation causing net water loss to basin system
- Bank erosion and silting
- Water acts as reservoir for effluents; these become concentrated if evaporation high
- Relocation of settlement if required
- Change to land use patterns
- Possibility of local earthquakes via lubrication of joints at dam itself
- Water overflows from spillway below dam
- Loss of nutrients to low-lying land downstream as silt levels fall.

(ibid., pp. 295–6)

To this list we can add erosion at the front of the Nile delta caused by the Aswan Dam, as sediment deposition is much reduced (Walker, 1997, p. 11), and salinisation on valley terraces as the water table changes following flooding or in delta areas as saline ocean waters become more invasive due to reduced runoff levels (Thomas and Middleton, 1997, pp. 73–4). We can also note that Meade 'strongly agreed' that earthquakes had been induced by several reservoirs in his study but 'strongly disagreed' in the case of others (1997). Goudie and Viles (1997) provide further details, plus a brief case study of the Colorado River, dammed by the Hoover and Grand Canyon dams among others.

---

dissidents. However, opponents of the Sanxia project would have seen themselves as identifying a complete misdirection of energy and resources, one that is difficult to sustain and maintain. The costs are too high, in diverse ways, as they are with other 'top–down' projects in China and elsewhere. In contrast they advocate, a more varied, 'bottom–up' approach to floods and related issues across the entire length and breadth of the Yangtze River basin.

Table 6.1 Factors in favour of the Sanxia project

Dimension	Factors
Economic	• Fits modernisation route to development • Is growth-oriented and concerned with high technology • Shows effective use of financial resources • Attracts foreign investment and expertise.
Political	• Illustrates power of the state and the CPC • Centralist solution to major problem.
Historical	• In line with large-scale interventionist traditions such as construction of the Great Wall or Grand Canal • Creates new 'wonder of the world'.
Spatial	• Concentration of resources into a single location as key solution to perennial flood problem.
Social	• Greater good of the population of the lower Yangtze, especially in the cities • Promotes consumerism via energy production and river transport of bulk goods.
Environmental	• Technocratic control over the environment • Illustrates might of humanity to overcome environmental constraints.

This would ideally involve full citizen participation, expenditure on environmental education, and a wide range of measures from the river uplands right through to the river mouth. The emphasis would be on *tian ren he yi* (see Chapter 2), oneness with nature rather than separation from it.

To a certain extent, some of the necessary analysis and change of attitudes required for a shift in policy became evident during and after the severe 1998 floods, when a number of articles in *China Daily*, *Beijing Review* and the *South China Morning Post* (SCMP) focused in detail on their causes. The official line, as evidenced by these sources, was variable but tended to concentrate upon such themes as ecological degradation, drawing on briefings from the State Environmental Protection Administration (SEPA) (*China Daily*, 13 August 1998), gross environmental abuse during the Maoist era according to one official, Zhuang Guotai of SEPA (SCMP, 11 August 1998), corruption according to local people in one affected area, as reported by *Agence France Presse* in the SCMP (30 July 1998), and 'stingy' officials who refuse to spend enough on conservation measures (SCMP, 28 August 1998). In brief, the key points of each of these are:

• **Ecological degradation.** Human activities – unplanned land cultivation, irrational construction of small water conservation projects, and excessive logging – have caused serious ecological degradation that impaired the function of lakes and wetlands in storing water and regulating flooding. The number of freshwater lakes in the areas around the central and lower reaches of the Yangtze fell from 1066 in the 1950s

Table 6.2 Factors against the Sanxia project

Dimension	Factors
Economic	• Encourages unsustainable development • Locks China into further highly wasteful expenditure • Financial resources could be better employed in other ways • Sheer scale of expenditure encourages corruption.
Political	• Needs citizen participation and grassroots intervention to deal with severe flood issues.
Historical	• Such large-scale projects in the past had unforeseen negative side effects, contributing to today's major problems; also posed a perennial problem of maintenance and upkeep.
Spatial	• To deal with a major spatial problem of flooding requires spatially diffuse measures the length and breadth of the river basin.
Social	• Stimulates wasteful materialist lifestyles, and at cost of those directly affected by resettlement.
Environmental	• Technocratic control over the environment creates more problems than it solves (such as pesticide use, GMOs) • Humanity should seek oneness with the environment, not control over it • Upsets vital *qi* of the Three Gorges region and beyond.

to around 182 in the early 1990s, the water area being reduced by 46 per cent. Because of silt deposition, the water storage capacity of Dongting Lake has shrunk from 29.3 billion cu, m in 1949 to 17.8 billion cu. m. Some factories discharge industrial waste into the watercourses. The country has constructed 86,000 reservoirs, but 40 per cent have silted up. The frequency of flooding of Dongting Lake has increased from once every 41 years before the 1860s to the present rate of once every 5 years. SEPA called on local governments to prioritise ecological preservation in flood control (*China Daily*, 13 August1998).

• **Environmental abuse in the Maoist era**. This includes massive destruction of forestry during the Great Leap Forward (see Chapter 3). 'The Yangtze Valley used to be pretty much covered by forest, but since the massive "make steel movement" in the 1950s, much of it was cut down', said Zhuang Guotai. The deforestation problem was especially serious in the upper reaches of the Jinsha, Min and Jialing rivers. 'The damage to this vegetation cover has led to soil erosion, mud and sand washing down the river and raising the river bed', Mr Zhuang said. 'Therefore, a not very big flow can still create a very high level of water'. Zhuang also referred to shrinkage in lakes, with the size of eight major lakes dwindling by a third from 1949 to the early 1980s, meaning that floodwaters enter directly into the Yangtze River. A further problem had been the development of land once reserved for water channels, as could be seen in Changhou town in Jiangxi province. 'Most of the land for farming and residence used to be the riverbed; now it is like a big island in the river. 'It has become a big tumour which blocks the smooth flow of the river', Mr Zhuang said (*SCMP*, 11 August 1998).

- **Official corruption**. Officials blame the failure to build a stronger anti-flood system on a lack of funds or on siltation problems, but local farmers accuse corrupt officials who, they claim, siphoned off money allocated for strengthening dykes. Many dykes are made only of earth, and while the local official view is that 'Every year we raise the level of the dykes but every year it becomes too low because of the silting'. A 'Hundred Boats' project had therefore been launched to dredge the riverbed between Wuhan, in Hubei province, and Jiujiang, to complement work to raise dykes. In Jiujiang, however, farmers accused officials of siphoning off flood-relief funds. A woman whose home in Hukou has been flooded since 24 June said: 'We do not want the money for ourselves. We want it put into a special fund to build a strong dyke so that we will not suffer such flooding in future.' A farmer living near the Yikung dyke said 400,000 yuan was earmarked for it a few years ago. 'Some work was done but not much, otherwise the dyke would not be in the shape it is today', he said. 'We do not know where all the money has gone' (*SCMP*, 30 July 1998).

- **'Stingy' officials**. Members of the National People's Congress standing committee who had been briefed on the floods previously by Vice-Premier Wen Jiabao criticised officials for their failure to limit the devastation. Zhang Huaixi, a former vice-governor of Jiangsu province, blasted officials for turning a blind eye to the problem until it was too late. '[Officials] would dump anything in the water when the floodwaters struck. But when the water recedes, they are so stingy that they won't put a thing into it', the China News Service quoted him as saying. 'There is no time to waste; we need to invest more on conservation facilities.' Technology specialist Wang Xuan also criticised the 'rotten perform-ance' of some officials. He said some cadres visiting flood-stricken areas had sought special comforts, asking to be housed in air-conditioned rooms and fed soft-shelled turtles. 'We should not tolerate such officials', Mr Wang said. Committee members supported the Govern-ment's decision to ban logging in Sichuan province (see below). Jiang Chunyun, an NPC vice-chairman and a former vice-premier in charge of agriculture, said the summer floods had served as a wake-up call to many people. It was time for China to rethink its environmental policy, he declared (*SCMP*, 28 August 1998).

Such issues as these led the government to announce in late August 1998, in a briefing to *Beijing Review* journalists, that all lakes which had previously been filled in to make fields will be restored, all forests which had been destroyed to open up wasteland will be replanted, and restrictions had been put on reclaiming and cutting forests along the Yangtze River. Flood relief measures for victims were also announced. The tone was self-congratulatory and upbeat, for 'the surging floods, which crested time and again, have been held back by a strenuous flood-fighting efforts involving millions of soldiers and civilians':

> The PLA made outstanding contributions. From the beginning of flood season until 24 August , the army, navy, air force and armed police forces had sent out more than 4.33 million troops, dispatched 236,800 vehicles and 35,700 boats and vessels, and flown 1289 aircraft and helicopter sorties. This has been the most massive military contingent ever mobilized against a natural disaster since the founding of New China in 1949. They helped rescue and evacuate nearly 4.2 million civilians, urgently repaired and strengthened 7619.16 km of dikes, plugged 5762 breaches and potentially dangerous leaks, and moved out 78.92 million tons of goods and materials.
>
> (Li Rongxia, 1998)

Such a tone contrasted markedly with suspicions of foreign journalists and others who were refused entry to the worst hit areas, that the death toll was many thousands and that the cities were saved only because dykes were deliberately blown up to flood large areas of farmland. The aim of the heroic propaganda was 'to evoke a sense of national unity in the face of the crisis and encourage people to be generous with their donations, so that it is not the government alone that is left with the burden of feeding and caring for the victims and rebuilding their homes and livelihoods' (*SCMP* 24 August 1998). This appeal to patriotism and fellow-feeling was successful, and a telethon was organised by the China Charity Fund to raise 600 million yuan, while billionaire Li Ka-shing pledged $50 million (ibid.). However, what was missing in all this, according to this highly regarded newspaper, was an analysis of the deeper causes of the flood – deforestation of the upper reaches of the rivers, leading to earth and sand flowing into the riverbeds, and the filling in of rivers and lakes to grow grain – that would lead to criticism of official policy.

Nonetheless, Water Resources Ministry officials were reported a few days later as claiming that flood-control facilities in 70 per cent of the mainland's 600 plus flood-prone cities were 'substandard'. An official said the government budget for infrastructure to control annual flooding had been disproportionately small in recent years. State investment grew at an average rate of 21 per cent annually from 1991 to last year, but totalled only 115 billion yuan, he stated, citing State Development Planning Commission figures. The figure represents just one per cent of infrastructure spending in the six-year period and is far outstripped by direct losses from this year's floods alone. (*SCMP*, 27 August 1998).

Another arm of government policy was the banning of logging in Sichuan, Yunnan and other locations along the Yangtze, recognising that, as Premier Zhu Rongji stated 'The disaster has, at least in part, been due to rampant deforestation which has led to serious erosion' (*SCMP*, 3 September 1998). As with other matters Chinese, the scale of this ban is huge. In Sichuan, for example, 54 counties in the Chuanxi forest area would be affected and all timber markets shut down, according to the Governor, Mr Song who announced the ban for that province (*SCMP*, 24 August 98). About nine million hectares of hillside was to be sealed off under a reforestation programme for 'livestock grazing and fuel gathering'. A report in *Xinhua* said that over 70 lumber firms were operating in the Chuanxi forest area but did not say how many workers were employed. However, it said the workers would be hired to be 'responsible for planting and protecting trees' (ibid.). Similarly, in Yunnan province logging was banned along Jinshajiang, a branch of the Yangtze River, beginning 1 September 1998 (*China Daily*, 2 September 1998). The latter source also suggested that Yunnan Province had over ten years already spent 170 million yuan (US$20 million) to save the forests, as a result of 'The Project of Yangtze Upper-Stream Shelter-Forest System', involving more than 667,000 hectares of shelter-forests planted to conserve water and soil along the Jinshajiang River. Although the evidence of previous chapters shows that such grand claims must be taken with the proverbial pinch of salt, what was clear was that Zhu Rongji was sayiang that Beijing would rather sacrifice its timber industry than allow rampant logging to continue. 'We can make up the shortfall by imports after we cut our timber production', Mr Zhu said. 'Imports are often cheaper than our own output. But we must make sure the timber workers' basic livelihood is looked after' (*SCMP*, 3 September 1998, op. cit.).

Measures such as these demonstrate a more realistic, holistic response to the manifold problems along the Yangtze Basin. By April 1999, however, a 32-page official report had reverted to blaming long-term climatic changes and exceptional conditions for the 1998 floods. In a press briefing Zhou Wenzhi, vice-minister of water resources, summarised the report as identifying as the major cause of the floods along the Yangtze River in Central China and the Songhua River in the north-east 'the abnormal climate last year' (*China Daily*, 28 April, 1999, op. cit.):

Climate changes included the prolonged effects of the El Nino phenomenon in 1997–98, the strongest this century; deep snows accumulated on the Qinghai–Tibet Plateau in the previous winter and spring; and excessive rainfall caused by subtropical high pressure that blocked a rain belt over the Yangtze. Climate changes such as these and the persistence of the rain belt had been a rarity in the previous 40 years, the report from the Ministry of Water Resources concluded.

(Ibid.)

While we do not doubt that some 'exceptional' conditions did occur that year, there have surely been too many regular floods along the Yangtze and elsewhere for all of them to be caused by 'exceptional' factors. Although the report concluded by referring to such measures as erosion control, a logging ban and the return of farmland to lakes there is a danger that over concentration on abnormal conditions allows the authorities to get off the hook of responsibility for a deep and serious response to the environmental crisis of the Yangtze Basin. As yet, no-one has had the courage to put the Sanxia dam on ice, but at least the attention being paid to such measures as those listed above illustrates that the dam itself cannot be the complete answer even if all its associated dangers can be overcome. These proposed measures themselves seem to us overly top–down, and not sufficiently involving the people themselves. The history of the PRC is full of examples of Beijing setting policies and the people blithely ignoring these in their own struggle for survival. As regards the dam, perhaps it is already too late for it to be stopped and it remains to be seen whether there can, even at this late stage, be a change of heart by the Chinese government. Perhaps the fears of critics, now also including ourselves, will turn out to be misguided, and the Sanxia project will become a shining example of humanity's urge to improve its environment, and of the strength rather than weakness of China's decision-making process. It will be dreadful indeed if the doubters are proved right, however, for the price to be paid by the people and environment of the Yangtze valley and China generally will be enormous.

7 Moving the waters

In Chapter 3 we considered some of the grandiose industrial and agricultural schemes prevailing during the Maoist era that had disastrous environmental results, under the general theme of 'politics in command'; in the last chapter we considered a latter day example in the shape of the Three Gorges Dam – although it is too early to tell if it, too, will also adversely affect the fragile mainland ecological balance. In this chapter, the theme is taken up again through a detailed examination of another dream of Mao Zedong – a massive transfer of waters from the lush lands of southern China to drought-prone areas in the arid north. We consider the various schemes that have been put forward, the reasons for them, their technological feasibility, and their environmental implications. This will lead into a detailed examination of the whole issue of water in the Chinese context – notably the difficulties of providing clean water, or, in a growing number of instances in urban areas of northern China, the problem of providing sufficient water of any description for household and industrial use.

Of the five major river basins that dominate the mainland, and in which 900 million of China's 1.3 billion people live, only the Yangtze currently has abundant water supply. The four northern basins – the Yellow, Hai, Huai, and Liao – all face acute water scarcity and a swelling diversion of water to non-farm uses. The Hai basin, which is home to 92 million people and includes both Beijing and Tianjin, is now in chronic deficit. The projected water withdrawals in the basin in the year 2000, estimated at 55 billion cu. m far exceed the sustainable supply of 34 billion cu. m, according to various government projections.

This water deficit of 21 billion cu. m can be satisfied only by groundwater mining. But once the aquifer is depleted, water pumping will drop to the sustainable yield of the aquifer, cutting the water supply by nearly 40 per cent. At a minimum, this indicates that the reallocation of irrigation water to cities, already underway in the region surrounding Beijing, will become basin-wide in the years ahead.

Yellow River – continued source of sorrow

The Yellow River, as noted in Chapter 2, is both the great cradle of Chinese civilisation and also its sorrow. Historically, the Chinese have always worried about its flooding. Records indicate between 206 BC and 1949 the banks collapsed in the lower reaches of the Yellow River more than 1,500 times, claiming millions of lives. At the same time, there is the worry over the heavy volume of mud it carries down to the sea. Each year, the river displaces up to 1.6 billion tons of soil, leaving much sediment on the riverbed that therefore constantly rises. Through protective dykes built along some stretches to cope with this phenomenon, the river level has risen so much that it now flows many metres above the surrounding land.

But now, the age-old fear of flooding has been replaced by a fresh concern: that the river is dying, threatening millions of hectares of crops. The river ran dry in its lower reaches in February 1997, earlier in the year than ever previously recorded. In 1996, the drought lasted a record 136 days. Since 1972, the river has run short of water along its 786 km lower reaches 20 times. The situation became especially acute in the 1990s (see Box 7.1 for an update in early 2000), as droughts began earlier and affected increasingly wider regions. This worrying trend prompted some experts to predict the river might ultimately become a continental river, one that never reaches the sea.

In 1997, *China Daily* reported on some of the implications of this, noting that 'in Shandong, a large grain producer, the river has been dry for more than 130 days, threatening 7.4 million ha of crops and drinking water supplies for 52 million people. A brief rainy spell last week caused the lower reaches to flow, but it stopped again after only 56 hours and was insufficient to affect agriculture' (*China Daily*, 12 August 1997). Shandong, it is worth remarking, is a crucial grain producing area. Failure of its crop due to lack of water would have severe implications for the nation's ability to feed itself, and send ripples around the entire world grain market.

According to Ren Guangzhou, senior engineer and deputy director of the Department of Water Resources Administration and Policy at the Ministry of Water Resources (author interview August 1997), the drought tendency can be traced to four causes. First, rainfall in the Yellow River Basin has been on the decline, especially in the 1990s. Meanwhile, demand for water has been increasing through rapid economic development. To prepare for the dry spring season – the sowing season that demands a lot of water – regions have started diverting water from the river into their reservoirs even in winter. As a result, the river carried only 8.8 billion cu. m of water between January and July 1997, down 50 per cent from previous years.

'Decreasing rainfall is not the main reason for the river drying', insists Ren. 'We have yet to work out a mechanism to effectively manage the water resources of the Yellow River.' It is true that in 1987, the central government, together with the ten provinces, autonomous regions and

Tianjin Municipality that lie along the Yellow River, mapped out a distribution plan for the river's available water runoff (rain flow from the land into the river). Precautionary measures were taken to guarantee each year at least 20 billion cu.m. of water got to the sea after diversion – in part to reduce the silt deposits on the river bed. But the plan failed to address distribution of water during severe droughts. Each year, as each region takes its annual share of the water, problems arise if they all rush to take water at the same time.

'The third cause of the problem is the rampant waste of water, whether it occurs in the upper, middle or lower reaches of the river', the official said. Nearly 90 per cent of the water diverted is for irrigation. Outdated irrigation facilities and methods worsen the waste. In Ningxia and Inner Mongolia, one hectare of land uses up to 15 tons of water, three times higher than the standard amount.

The fourth cause of the problem is insufficient flood control facilities in the middle reaches of the river. 'Sixty per cent of the river's annual runoff is concentrated in July, August and September. The figure for April, May and June, the traditional peak seasons for water use, only accounts for less than one-fifth of its annual runoff.' If there were enough reservoirs in the middle reaches to store water in the flood seasons for the dry seasons, then the problem of the lower reaches would be greatly reduced. In the upper reaches, there are several large reservoirs, but they can hardly help regions in the lower reaches that are too far away.

Water diversion

The Yellow River problem has thus helped to revive interest in a plan first mooted in the Maoist era to divert the abundant water resources available in south China to irrigate the arid farmland in the north. Essentially, the idea is to draw off water from the Yangtze – up to 70 billion cu. m a year – and channel it through one of three possible routes to the Yellow River. Mao Zedong first proposed the idea of transferring water from one river to the other in 1952, and over the ensuring half century, some 90 expert reports have been produced on aspects of hydrology, geology, seismology, environment impact and cost. The Ministry of Water Resources revived the idea in the mid 1990s, and work was due to begin on the first phases of the project as this book was written.

The aim of the project is to overcome a natural imbalance by which the Yangtze has what is officially considered more than sufficient water to meet existing needs – and needs projected into the third decade of the twenty-first century – while the Yellow River far too little. In late spring 1995, for example, water in the Yangtze ran high above its danger line, inundating vast areas of agricultural land and causing heavy losses. At the same time, some areas of the Yellow River were reduced to a tiny trickle. The same thing happened in 1998.

Box 7.1 Crushing drought robs millions of running water

By May 2000, what was described as 'the worst drought in more than a decade' had reportedly left 15.5 million people without running water in China's central plains. The Yellow River had dwindled to no more than a trickle from Henan province to its mouth in Shandong province, leaving millions of hectares of farmland without irrigation. In Henan, total rainfall for the previous 12 months was only 20.3 mm, 14 per cent of the yearly average. About 979 of the province's 2272 reservoirs were dry and wheat production had fallen more than 50 per cent from the average, local officials reported. The water table fell between two and four metres in the spring alone. Hongze Lake, the biggest of several large lakes in northern Jiangsu, covered only 63 per cent of its basin, and neighbouring Weishan Lake 46 per cent. Most smaller lakes and rivers in the region were completely dry.

Although rain had been scarce over the previous 20 months, this alone does not explain the depletion of the region's water resources. Agriculture accounts for more than 80 per cent of water use in central China. The area is heavily dependent on irrigation. Rainwater is rarely enough. Redirecting water from rivers or pumping it from lakes and underground sources goes on unrestricted. And water as a public utility is largely subsidised, giving users little incentive to reduce consumption to a sustainable level. In Weihai, Shandong province, the government began implementing policies to curb the use of water more than a year ago, raising the price of water from 80 fen per cu. m to 1.2 yuan. But local officials said this had had little effect, so they took drastic measures by raising the price to 40 yuan. Locals responded by re-using the same water to wash themselves, vegetables, clothes, floors and toilet bowls, the Guangzhou-based *Nanfang Weekly* reported.

Some specialists hoped the drought would impress upon the Government the need to change its water resource management policies permanently. Without adequate incentives to change their practices, farmers persisted in using wasteful methods. Some farmers continued to plant rice, using two tonnes of water to produce one kilogram of grain, right up until the pipes ran dry, the *Nanfang Weekly* reported.

In Jiangsu province, 1.13 million ha were drying up, hindering spring ploughing, the *Yangtze Evening News* reported. In the northern province of Shaanxi, the drought was affecting some 667,000 ha, 40 per cent of the province's farmland, Xinhua news agency said. In Ankang prefecture, in the south of the province, 60 per cent of farmland had been hit hard, with 170,000 people affected in what the report called the worst drought on record. In Huoqiu county, Anhui, 66,000 ha of farmland were seriously affected by drought as water levels in the county's reservoirs measured only 59 million cu. m, about one-fifth of reserves held at the same time in the previous year, the *Farmer's Daily* said. Meanwhile, the western part of Guangdong province was experiencing its worst drought since 1954, the *Yangcheng Evening News* reported.

The first five months of 2000 were reportedly the driest in Beijing since 1949, and the previous year's total rainfall was only 58.7 per cent of the average. The water table under the capital dropped 2.3 metres from the beginning of the year until late May, adding up to a fall of more than 60 metres over a 35-year period.

In 2001, there were more problems around the country. A withering drought that began in Sichuan province in October and persisted throughout the ensuing winter and spring was described as the 'worst in 30 years' causing severe damage to power supply (21DNN.com, 5 April 2001). Persistent drought in north-western and northern regions affected 23 million ha of farmland, with 8.9 million ha producing no crops at all, making a good harvest 'a near-impossible hope' (*China Daily*, 7 June 2001).

The impact of China's dual problem of water scarcity and water pollution exacts a costly toll on productivity. Chinese residents currently face a shortage of 28.8 million cu. m. of water daily. According to Vaclav Smil these shortages cost the Chinese economy between 5 and 8.7 billion yuan (US$620–1.06 billion) in 1990 (Smil, 1995). Another authority suggests that water shortages in cities cause a loss of an estimated 120 billion yuan (US$11.2 billion) in industrial output each year (World Bank, 1997, p. 23).

Meanwhile, population and water use per capita are growing; the physical condition of China's water facilities is ageing; competition between the potential uses for water is increasing; aquifers are becoming depleted; water pollution is rising; and the social cost of subsidising increased water usage is increasing. China has a total of 2800 billion cu. m of annually renewed fresh water; the world's most populous country is fourth in the world in terms of total water resources (WRI, 1999). Considering per capita water resources, China has the second lowest per capita water resources in the world, less than one third the world average. Northern China is especially water-poor, with only 750 cu. m per capita; this geographic region has one-fifth the per capita water resources of southern China and just 10 per cent of the world average (World Bank, op. cit., p. 88). Beijing's per capita water resources, meanwhile, are even worse – an eighth of the national average, one-thirtieth of the world average, and even lower than Israel's (*China Youth Daily*, 13 November 1999).

Box 7.2 Implications of the water shortage

China's rapid economic growth, industrialisation, and urbanisation –
accompanied by inadequate infrastructure investment and management capacity – have all contributed to widespread problems of water
scarcity and water pollution. Of the 640 major cities in China, more
than 300 face water shortages, with 100 facing severe scarcities (NEPA,
1997). Experts at the China Academy of Sciences (CAS) in 1998
warned that the problem in the countryside was even more severe.
More than half the country's peasants lacked sanitary water to drink,
with 50 million farmers and at least 30 million farm animals facing
severe water shortages, they warned (*Economic Information Daily*,
11 August 1998). Two years later, Minister for Water Resources Wang
Shuheng was warning of severe water shortages by 2030, when the per
capita share of resources is expected to have dropped as much as 20 per
cent given the inevitable population growth. The shortages were being
exacerbated by wastage of irrigation water, which had reached about
60 per cent compared with 30 per cent in the rest of the world, he said
(*People's Daily* 25 March 2000).

The minister sketched the following grim scenario. By the middle
of the century, China's gross domestic product is expected to increase
ten-fold, requiring more water in urban areas and for industries. At that
time, its water resources will have to meet the demand from 140 million
tons of grains and an urbanisation ratio of 70 per cent, up from the
present 40 per cent. Annually, more than 26 million ha of farmland, or
over one-fifth of China's total, is plagued by drought. The shortage of
water is worsening in both rural and urban areas. In irrigated areas,
the annual shortage of irrigation water has reached 30 billion cu. m.
Meanwhile, a serious waste of water continues. Only 40 per cent of the
irrigation water has been used effectively, compared to 70 to 80 per cent
in other countries. In China, 10,000 yuan worth of industrial output
requires 103 cu. m of water, 10 to 20 times more than in developed
countries. Although China's per capita share of water resources is
only one quarter of the world's average, only about half of water is
recycled, 35 per cent lower than that of developed countries (*China
Daily* 23 May 2000).

As discharges of both domestic and industrial effluents have
increased, clean water has become increasingly scarce. According to the
CAS experts, coastal cities such as Tianjin and Shanghai were finding it
increasingly difficult to provide residents with clean drinking water.
Seawater contamination had become common, made worse because
of insufficient measures taken by companies to ensure the proper
dumping of waste materials. Soil erosion in the north-west has also
taken its toll on clean water supplies (ibid.).

The distribution of groundwater is similarly skewed: average groundwater resources in the south are more than four times greater than in the north. Dramatic shifts in annual and monthly precipitation cause floods and droughts, which further threaten economic growth. As surface water quality has worsened, the Chinese have increased their extraction of groundwater to meet water demand. As a result, over-extraction has become a serious problem in a number of cities including Nanjing, Taiyuan, Shijiazhuang, and Xi'an. Groundwater depletion is most problematic in coastal cities, including Dalian, Qingdao, Yantai, and Beihai, where saltwater intrusion is on the rise (NEPA, op. cit.). Although there is no comprehensive monitoring of China's groundwater, some studies suggest its quality, not just quantity, is severely threatened in many regions. According to one estimate made in the mid 1990s, half the groundwater in Chinese cities has been contaminated (Zhang *et al.*, 1994, pp. 215–17).

The water shortage also has socio-political implications. A China case study by the US Council on Foreign Relations suggested that, in the long term, water scarcity could significantly diminish the state's capacity to maintain control. It said both demand- and supply-induced water scarcity had already resulted in substantial inter-provincial conflict. Continued population growth, as well as increasing demands from industry and agriculture, cimbine to diminish the coherence of the state by engendering a growing number of inter-provincial claims to these water resources. Rising pollution levels also result in growing inter-provincial disputes over the responsibility and costs of treatment facilities and clean-up costs. These problems, the council believed, were endemic, with little prospect for immediate resolution. Moreover, the central government had yet to develop an effective mechanism for resolving such conflicts (Economy, 1998). And time may be running out. Another CAS 1998 report predicted water usage would rise 60 per cent to 800 billion cu. m by the middle of the current century from the 500 billion cu. m level prevailing in the late 1990s. 'This would put annual demand at 28 per cent of usable water reserves. Experience tells us a water crisis will arise when demand exceeds 20 per cent of reserves', the report said (*Xinhua*, 4 June 1998).

The original water diversion project plan identified three separate routes. A 'western' route would start from the upper reaches of the Yangtze, run through the Qinghai–Tibet Plateau and end up in the upper reaches of the Yellow River. This takes advantage of the fact that at this point the two rivers are only a few hundred kilometres apart, before diverging and reaching the sea very far apart. A 'central' route would have water diverted from the northern-most point of the Han Shui River, a Yangtze tributary, directly to Beijing. The third or 'eastern' route would divert water as the Yangtze approaches Shanghai, sending it north to Tianjin, the large industrial city located roughly 120 miles from Beijing, both chronically water-short cities (see Boxes 7.1 and 7.2), relying heavily on the existing channel of the 1,500 year-old Grand Canal.

The latter route is the simplest to achieve, but would benefit only relatively small areas along the eastern coast that already enjoy great prosperity, and is likely to further exacerbate the east–west economic gap. The central route is regarded as posing few serious technological challenges, but again has its shortcomings in regard to the areas that would benefit. Only the western route would benefit the whole course of the Yellow River, as well as serving an additional purpose of helping to open up the economically-backward, but resource-rich Qinghai–Tibet plateau and other poverty-stricken areas in the far west. The advantage of this route is the close proximity of the upper tributaries of the two rivers. But the technical challenge is immense. Differences in the depths of the riverbeds would mean that diversion could only be achieved through the construction of a 300-metre high dam. This, in itself, is no easy task in a frigid zone at an altitude of 3000–4000 metres. Then extensive tunnels would also have to be driven through the permafrost that makes up the watershed of the Bayanhar Mountains. At the time of writing, the Ministry of Water Resources had approved work to begin on the eastern and central routes, while more study was said to be necessary before attempting the western channel.

More immediately, at the end of 1999, plans were submitted to the central government for a 54.8 billion yuan 'new Grand Canal', with engineers estimating the 1246 km waterway will take six years to build, linking central Hubei province's Danjiangkou reservoir with the suburbs of Beijing. The construction would be on a par with the existing Grand Canal. That 1789 km waterway, the longest in the world, winds south from Beijing through Tianjin and Shandong Province down to Anhui and Jiangsu provinces, and ends in Hangzhou, capital of Zhejiang Province. It has played a crucial role in the nation's transportation system for almost two millennia.

But distorted navigation routes, shallow water and low bridges have hampered the canal's handling capacity in recent years, and only the section south of Shandong Province is now able to accommodate water traffic. What is worse, it is now seriously polluted as a result of mushrooming industrial development in Jiangsu and Zhejiang provinces. Sewage treatment plants, however, are now being introduced to clean the canal. Zhejiang Province alone was building 20 sewage disposal plants at the time of writing. Once-reluctant enterprises along the canal are also now willing to pay for waste treatment because doing so has become a prerequisite to getting further official loans. A proposal has been approved to connect the canal with one that runs from Hangzhou to Ningbo, a coastal city in Zhejiang Province, to dilute the polluted water.

As this book was being written, work had begun on dredging and widening the waterway to enable it once again to play a crucial role in the movement of people and goods. In addition, renovation will allow the diversion of water to traditionally parched areas of Shandong Province.

As far as the 'new Grand Canal' is concerned, an expert assessment suggests it would bring 14.5 billion cu. m of water annually to Beijing, whose exploding population (see Chapter 5) is placing impossible strains on available ground water – the capital's water table having sunk more than three metres on average during the 1990s. Added to this are increasingly frequent prolonged droughts, part of a climate change in the northern part of the country that has seen the rapid expansion of desert regions to some extent due to centuries of deforestation. The drought that has prevailed almost throughout the 1990s has left riverbeds dry and water levels in reservoirs at record lows. The 1999 rainfall during the traditional summer rainy season was negligible.

However, let us leave the tranquil Grand Canal for now, and return to the turbulent waters of the Yellow River.

Not enough water

A cradle and a killer, a mother and a menace: from time immemorial, China has both depended on and dreaded its mighty rivers. During the Qing Dynasty (1644–1911), control of the Yellow River was thought so critical to dynastic stability that a ministerial post, Governor of Yellow River Affairs, was established. The holder, only second in power to the prime minister, could enter the Forbidden City without dismounting. But in the event of floods, heads would roll. The precedent of history lends an ominous warning to today's officials trying to deal with the petulant waterway.

Since 1949 some eight dams have been built and four more were under construction at the time of writing along its 4674 kilometre course to control flooding, among other purposes. It is somewhat ironic, therefore, to find officials now worrying about finding enough water for the Yellow River to find its way out to the sea.

'I was shocked to see that during the usual 90-day flood season from July to September in 1997, a total of 72 dry days were recorded at the Lijin hydrometric station, the last such facility before the Yellow River empties into sea', recalls Chen Xiaoguo, chief engineer and deputy director of the Yellow River Conservancy Commission based in Zhengzhou, capital of Henan Province (author interview). During 1997, he said, Lijin in Shandong Province recorded 'zero volume' for 226 days – the worst drought since records have been kept. According to the commission's research, the river should carry 20 billion cu. m of water into the sea annually. In 1998, it managed only 1.8 billion cu. m.

Chen, who has studied the river since 1962, says that until recent times, the river posed a year-round threat: 'In summer it flooded of course, but the thaw of its ice also caused havoc every spring with ice runs'. According to him, in the 25 centuries between 602 BC, the fifth year in the reign of King Ding of the Zhou Dynasty, and 1938, the Yellow River breached its dykes no fewer than 1,590 times. In the 50 years between 1896 and 1946, there were 210 breaches. In addition, Chen points out, the river's course to the sea in its lower reaches has shifted from time to time because of complicated tectonic structures and movements under the Yellow River basin, as well as breaches.

'Until recently there were on average two breaches every three years, and a major change of course once a century', he says. 'Each and every one of these events left its toll on local populations, both their lives and property.' Each inundation has also left a layer of mud, the main culprit of flooding in the first place. For in its middle reaches the big bend of the river passes through an easily eroded plateau of the fine sediment called loess. When it rains, flash floods carry mountains of loess down into the river. The further downstream one goes, the muddier the Yellow River becomes. The slower it flows, the more sandbanks it deposits along its course.

However, some of this sediment, as well as that found in the Yangtze River, comes from as far away as the Qinghai Plateau where they both originate, and is a source of deep concern for provincial government officials there. According to provincial Vice-Governor Liu Guanghe, an area of 33.40 million ha, or 46 per cent of Qinghai's territory, now suffers from serious soil erosion, resulting, on average, in 115 million tons of soil and sand being washed from the area into the two rivers annually, contributing to an average of 133,000 ha of arable land in Qinghai being turned into desert every year. This erosion is partly due to drought, but widespread deforestation in previous decades that have reduced the province's forest coverage to 2.3 per cent, according to Kang Weixin, director of the provincial Environmental Protection Bureau. The millions of tons of sand and silt, washed downstream by the river flow, pose serious threats to the normal operation of water control facilities and hydropower stations located along the courses of the two 'mother rivers', while the amount of water flowing down the Yellow River in Qinghai in the late 1990s was 23 per cent less than that in the 1970's (*China Daily* 19 April 1999).

Over the years, says Chen Xiaoguo, sedimentation has raised the riverbed by 5–10 centimetres a year, so that the riverbed in the lower now reaches 3–5 metres above the level of surrounding fields. This sedimentation also explains why the Yellow River has been of imperial concern for centuries, says E Jingping, director of the Yellow River Conservancy Commission (author interview). Floods have caused famines, and empty stomachs have toppled dynasties. 'During the Qing Dynasty, some governors of Yellow River Affairs were driven to suicide so as to save their families from disgrace and punishment', E says. And it is still a life or death job of sorts to this day, as E and his colleagues admit to the pressures of making sure that disasters are averted.

'We live in constant fear and endure enormous pressures', says Zhang Mingde, chief engineer with the Shandong Bureau of Yellow River Affairs under the Commission (author interview). 'For years I could not join my family at Chinese New Year – I had to be on guard against the ice run' (as this book was being written, heavy icing of the river was once again causing concern about massive floods with the spring thaw).

In fact the river has not breached its dykes for nearly half a century, which the commission regards as 'a remarkable achievement'. Chen estimates that keeping the Yellow River at bay has saved the country 400 billion yuan. 'The dams in the upper and middle reaches have effectively enhanced our ability to control floods and have adjusted the water flow so that ice runs could hardly form,' he says. 'Today problems are entirely different: there is not enough water.' The river dried up in part of its course during 21 of the 26 years up to 1998. And since 1991, its lower section has dried up every year.

'As the dry-ups have become more regular, both the number of dry days and the length of the river course afflicted have increased', says Zhang Mingde, who has been working on the Yellow River since 1961. 'In 1998, year its flow was 700 kilometres short of the sea.' But reduced water flow does not alleviate the burden of flood control. On the contrary, the threat of flooding increases. 'One of the fundamental causes of the Yellow River floods has been sedimentation', explains Zhang. 'In a normal year 1 billion tons of sand should be dumped into sea by the river water. Regardless of water flow or not, the sandbanks on the riverbed block the main course, and that means danger when flow resumes or increases.'

In August 1996, a flood hit the Henan-Shandong border area. 'At a flow volume of just 4,000 cu. m per second, the flood crest rose to a water level usually created by a volume of 10,000 cu. m per second – beyond the maximum designed capacity of our embankments', Zhang recalls. In other words, it takes less water now to make a flood happen. In 1998, the 1.8 billion cu. m of water that emptied into the sea carried only 200 million tons of sand, one fifth the normal amount.

The river's drying-up is partially due to unfavourable natural conditions, according to scientists studying the issue. Wu Chuanjun, a geographer and a member of the CAS, describes the Yellow River as 'congenitally deficient' since its drainage basin is mostly arid, with annual precipitation averaging 200–500 millimetres (author interview). As China's second longest river, its total annual runoff ranks only fourth against the Yangtze's 1,000 billion cu. m.

Weather factors compound the problem, says Tao Shiyan, an atmospheric physicist and CAS member (author interview). 'Climatic data at six meteorological stations in northern China indicate that annual precipitation in these areas has dropped dramatically since 1965, by 20–25 per cent from normal levels'. According to Tao, the great 700 km dry-up of 1997 was an extreme year, partly blamed on the El Nino phenomenon, which made an already thirsty Yellow River even drier. But he also points out that the 1997 drought was not the worst in history. During the 1920–40 period of prolonged drought, precipitation in the Yellow River basin plummeted by 35 per cent, with a mere 22 billion cu. m of runoff recorded in the worst year, compared with 25 billion cu. m in 1997. Yet the Yellow River never ran dry during those years. Therefore, Tao says, congenitally deficient water resources and unfavourable weather conditions 'are not entirely decisive factors' fully explaining the drying up of the Yellow River. 'They just play a role in it.'

After a 15-day inspection tour tracing the river – flowing and dry – in 1998, Tao, Wu and other scientists concluded that the major cause for the drying up of the river is 'excessive water use far beyond the river's capacity'.

Competition for water

Sixty per cent of the nation's farmland is north of the Yangtze River, where only 18 per cent of the country's total water resources are available. The Yellow River basin embodies the water dilemma confronting all of China.

'A 1995 survey ranked China 122nd in terms of water resources available to an average person among 153 countries and regions around the world, at 2300 cu. m per person every year', says Wang Hao, chief engineer with the Institute of Water Resources under the Ministry of Water Resources. 'That is about one-quarter of the world average. But the water resources available in the Yellow River basin are even scarcer, only one-quarter of China's average' (author interview).

In the Ningxia Hui Autonomous Region, annual precipitation is around 305 mm, but evaporation is as much as 2000 mm. People there historically depended on the Yellow River to irrigate their farmland for more than 2000 years. 'Without the Yellow River, there would be no Ningxia', asserts Mao Rubai, the local party chief (author interview). As almost the only waterway through the region (with a population of 5.28 million), the Yellow River accounts for 80 per cent of Ningxia's irrigation water supply. By the end of 1997, according to the region's water conservancy department, Ningxia's irrigated farmland increased from 310,000 ha in the early 1950s to 412,000 ha, while its grain output rose from 160,000 tons to 2.5 million tons. Meanwhile, a total of nearly 5 million ha of farmland in the nine provinces and autonomous regions along the Yellow River are now irrigated by its water, compared to 1.4 million ha in the early 1950s, according to Chen Xiaoguo. On the whole, water consumption by agriculture, industry and urban areas has more than doubled from 12.2 billion cu. m a year in 1950 to well over 30 billion cu. m in the 1990s.

'It is thought quite high if 30 per cent of a river's water resources are utilised', Chen says. 'In the case of the Yellow River, the figure is more than 50 per cent.' Liu Changming, a geographer and expert on water resources with the CAS Geography Institute estimates that of the more than 30 billion cu. m of the Yellow River water consumed every year, 92 per cent goes to irrigation, but only 30 per cent of that water effectively reaches the end crops, suggesting enormous waste.

Jia Dalin, an irrigation expert with the Chinese Academy of Agricultural Sciences, has found that with the same quantity of water, the Jinghui Canal Irrigation Area in Shaanxi Province benefits 80,000 ha, while the People's Victory Canal Irrigation Area in Henan Province benefits only 40,000 ha, the difference being explained by the leak-proof lining adopted in the former. His research found that of 122 diversion works employing Yellow River water for urban water supply and agricultural irrigation, only 10 per cent of the ditches had desirable seepage-proof linings.

And yet, despite the prolonged bouts of drying-up along the river's lower reaches, efforts continue to draw off even more water from further upstream. According to a 1998 study by the US-based Worldwatch Institute, demands on the river were expected to soar in the years ahead, because the Yellow River Basin, which contains 105 million people, had been designated by the central government for rapid industrialisation, so that each of the upstream provinces planned to increase its withdrawals from the river for residential and industrial uses.

Brown and Halweil, authors of the report, cite the example of a canal to be completed in 2003, to move 146 million cu. m of water per year into Hohhot, the capital of Inner Mongolia, to cope with the swelling residential needs of 1.2 million people as well as the demands of expanding industries, including the all-important wool textile industry supplied by the region's vast flocks of sheep. Another project, the Lijiaxia Hydropower Station, one of the largest in China, began operating its five 400,000 kW turbines in 1997. It is only the third of many large power stations scheduled for construction on the upper reaches of the Yellow River.

> Hydro-electric engineers like to argue that such plants merely take the energy out of the river, not water. But hydroelectric reservoirs, which greatly expand the river's surface area, can increase annual water loss through evaporation by easily ten per cent of the reservoir's volume. With the proliferating of new upstream projects, ever less water will flow to the already-depleted lower reaches of the basin. One result is that some companies are moving their factories upstream, both to assure an uninterrupted supply of water and to take advantage of the cheaper labour they can find there.
>
> (Brown and Halweil, 1998)

The two authors go on to say that in assessing the effect of future water losses on food production in China, it would be helpful to know how much of existing irrigated grain production was based on the unsustainable use of water, or groundwater mining.

They noted that, in the United States, where only one-tenth of the grain harvest comes from irrigated land, irrigation water losses would not substantially alter the world grain supply. But in a country where 70 per cent of an even larger grain harvest comes from irrigated land, and where ground-water mining is widespread, the impending consequences of aquifer depletion inevitably were much greater.

Another project, not dealt with by Brown and Halweil, that draws water from the Yellow River is the Wanjiazhai Water Transfer project, aimed at delivering 1.76 million cu. m per day from the Wanjiazhai Dam via a series of pumping stations to the booming city of Taiyuan in Shanxi province, 285 kilometres away. Again, growing population and industry have sucked out excessive amounts of ground water, while effluent from these people and activities has seriously polluted the local river. Qingtao *et al.* (1999), from the North China Institute of Water Conservancy and Hydropower in Zhengzhou, Henan Province, provide a detailed study of this project, but their environmental impact analysis does not deal with the issues raised by the Worldwatch Institute report.

Water shortages are not the exclusive problem of the north. In 1999, for example, residents of Shenzhen, known as the wettest of southern cities, had to queue up for water due to a serious drought. The problem was compounded by the lack of local large-scale reservoirs to catch the seasonal rainfall. At the time of writing, the city was about to embark on a project to divert clean water from the upper reaches of the Dongjiang River, a branch of the Pearl River. The first phase of the 105-km-long channel will direct water into the Songzikeng Reservoir, while a second phase will link up all reservoirs in the city's six districts. City government data indicated that Shenzhen residents had just one quarter of the national average amount of water per person, but according to the city's blueprint, it will have a population of five million by 2010, as against two million in the late 1990s. This would require a yearly water supply of 1.9 billion cu. m, while the existing maximum capacity was one billion, officials explained. (*China Daily*, 22 February 2000).

What are the alternatives?

Returning to the main issue considered in this chapter, Brown and Halweil have described the plan to divert water from the Yangtze tributary, the Han Shui, to Beijing as 'comparable in reach to turning to the Mississippi River to satisfy the needs of Washington, DC', with cost estimates soaring into the tens of billions of dollars for the benefit of a few. The western route, seen as bringing benefits to the widest area and the most needy, would be equally as expensive.

> Some analysts point out that money spent on south-to-north water diversion projects could be spent much more profitably on investing in water efficiency or importing grain. Those urging the latter point out that importing 20 million tons of grain per year (Canada's annual grain exports) to north China would free up the 20 billion tons of water that would be diverted by the Han Shui scheme, but at a much lower cost.
>
> (Ibid.)

At the same time, even diversion will not work unless China also tackles the demand side, through restructuring of the entire agricultural, energy, and industrial sectors to make them more water-efficient. This entails shifting to reliance on more water-efficient crops and livestock products and on less water-intensive energy sources. It will also mean reducing pollution so that water does not become unusable for irrigation (see Box 7.3 for a brief discussion on the pollution issue).

Along with more water-efficient agricultural techniques, there would be a need to shift to less water-intensive crops. This might mean producing less rice and more wheat in some regions. With livestock products, it would mean raising more poultry and less pork, since a kilogram of poultry requires only half as much grain, and therefore only half as much water, as a kilogram of pork – a staple of the Chinese dining table. And it might even mean an official policy of discouraging consumption of livestock products in the more affluent segments of Chinese society, where animal fat intake has already reached health-damaging levels.

Another option is to increase water use efficiency in homes and industry. For example, it might well make economic sense for cities in water-scarce regions of China to introduce composting toilets rather than the traditional water-flush toilets. The Western water-intensive sewage disposal model simply may not be appropriate for water-scarce China. Beyond this, the adoption of water-efficient standards for household taps and showers can also help stretch scarce water supplies. According to Brown and Halweil,

> [as] new cities rise throughout China and older cities expand and are rebuilt, urban planners would do well to keep the streams of industrial and residential wastewater separate – as opposed to replicating the Western model which combines these flows. Uncontaminated by industrial pollutants, residential wastewater can be recycled, while nutrients are removed for use as fertiliser. The present rush of expansion, while environmentally damaging and difficult to manage, at least offers a unique window of opportunity for efficient design, because poor designs adopted now will incur the economic costs of future retrofits and the social costs of water shortages.
>
> (Ibid.)

There is also much potential for saving water in industry. For example, the amount of water used to produce a ton of steel in China ranges from 23 to 56 cu. m, whereas in the highly industrialised countries, such as the United States, Japan, and Germany, the average is less than 6 cu.m. Similarly, a ton of paper produced in China typically requires at least 450 cu. m of water, whereas in industrial countries, it generally requires less than 200 cu. m.

For some industries, achieving high efficiency will require investing in entirely new technologies and factories. In other cases, rather modest changes in manufacturing processes can yield large water savings. In the energy industry, fundamental restructuring is already a global imperative because of the need for climate stabilisation. Fortunately, the technologies that offer the most immediate environmental benefits from the standpoint of greenhouse

---

Box 7.3 'Stupid deeds of ecological destruction'

According to a *People's Daily* reporter, writing about the Fourth National Environmental Convention held in Beijing in July 1996: 'In the past and the present, stupid deeds of environmental pollution and ecological destruction have occurred because some cadres, in particular cadres in leadership positions, are only concerned with developing the economy at the expense of the environment. The result is that more losses are incurred and grave consequences are irreversible.' (*People's Daily* 18 July 1996). River and lake pollution is a good example.

Statistical data made available during the conference showed the percentages of undrinkable water of category 4 and 5 in the river basins were respectively 67% for Songhua River and Liao River, 60% for Yellow River, 51% for Huai River, 41% for Hai River, 24% for Yangtze River, and 22% for Pearl River (*Ming Pao*, 16 July 1996). A Hong Kong reporter named Chen Guidi, touring 48 cities along the Huai River in 108 days, found that of its 191 larger tributaries, 80% of the water had turned black and stinky; two-thirds of the river had totally lost any use value (*Ming Pao*, 5 September, 1996). A three-year survey by the Water Works Bureau concluded there was no 'appropriate control of sewage' (*Xinhua News Agency*, 13 February 1997). The source of water pollution comes mostly from the dumping of factory wastes. Although, the authorities closed down almost 1,000 small-scale paper and leather factories along the Huai River, the damage had already been done. In 1996, the State Council ordered the closing down of 15 types of small factories regarded as highly polluting. In the beginning of 1997, 50,000 factories had been closed down (*Wen Hui Bao*, 13 January 1997).

In 1999, Guangzhou faced what was described as its 'most serious drinking water crisis in over 50 years' because of invading salty tides. The quality of the city's water source, the Pearl River, was worsening and the content of chloride in the drinking water far exceeded the state's safe drinking water standard of 250 mg per litre. The Xicun Waterworks, capable of handling 1 million cu. m of drinking water daily, had to cut production. Four other waterworks, capable of a combined water supply of more than 660,000 cu. m daily, followed suit, reducing the city's water supply by about 600,000 cu. m per day. According to local officials, the crisis stemmed from several months of scarce rainfall so that 'the waste poured into the Pearl River cannot be well diluted and algae have been propagating rather rapidly' (*China Daily*, 3 May 1999)

Pollution has turned tributaries of the Pearl River red and black (*China Environment Daily*, 27 April 1998). A survey by the environmental protection agencies of Guangxi and Guizhou provinces found the Honghu River, the largest tributary of the Pearl, turned black for several months a year and had been renamed the Heilongjiang, or 'black dragon' river. Tributaries were turning black in the low-water season and red in the flood season, and at both times the water was undrinkable. The most polluted river was the Tuochang Jiang, where the current could not carry away the amount of waste entering the water from local coalmines and processing factories. The study said the Hongshui River was turning red because increasing quantities of soil were being washed away during the rainy season along the upper reaches.

A spate of untreated sewage and illegal industrial waste discharges that triggered massive fish kills prompted the Ministry of Agriculture to issue an urgent circular in early 2000 demanding government agencies do a better job of halting the pollution of the nation's rivers and lakes. Sources in the ministry's Fisheries Bureau said a chemical fibre plant and a banknote paper mill in Baoding, Hebei Province, discharged untreated sewage into Baiyangdian Lake, contaminating killing fish valued at 23.8 million yuan. Anhui, Fujian and Liaoning provinces also recorded massive losses of fish, crabs, shrimps, and shellfish as a result of rampant water contamination. Investigation was difficult because most of the pollution-causing firms were major sources of revenue for local governments. (*China Daily*, 25 April 2000)

gas reduction – wind and solar power – are also water-efficient; they use much less water than hydropower, nuclear power or coal. Developing wind-power resources would also strengthen the economies of the wind-rich interior provinces.

Clearly, an across-the-board effort to restructure China's water economy is needed. In contrast to the traditional supply-side solutions to water scarcity, often involving gargantuan feats of engineering with adverse social and environmental effects, future water needs will have to be met by demand-side management instead and greater emphasis on per capita water efficiency.

In February 2000, the central government began drafting a long-term plan for controlling floods, cut-offs, pollution and the build-up of sediment on the Yellow River. A report by the Ministry of Water resources, setting two targets – tackling eco-environment problems such as erosion, and flood-control – estimated the cost at 144.1 billion yuan, of which 64 per cent would have to come from central coffers. In part, this could be met by raising the price of the river's water for agricultural and industrial use, suggested Chen Xiaoguo, deputy director of the ministry's Yellow River Conservancy Commission, but this would require the creation of a new regulatory agency to oversee ever-growing demand from the rapid development of north-western China. The ministry report said the government would have to limit water-consuming industrial projects and diversion-works.

Experts who drafted the report warned that 'the limited water resources of the Yellow River cannot be used for drinking or irrigation unless there are effective pollution controls'. The increased annual cut-off of the river's flow, caused by a worsening water shortage, greater demand for water, poorly regulated water supply, and by pollution, can be solved by adopting water-efficient irrigation and re-using 75 per cent of the industrial water supply. Diverting four billion cu. m would also alleviate the cut-off problem from the Yangtze each year, they declared (*China Daily*, 2 March 2000).

The report also expressed concern about the prospects for future flood disasters, pointing out the need to spend heavily on improving the quality of the defences. It claimed that the river had substandard dykes along a 440-km stretch of the lower reaches, while another 471-km section had dykes that leak (ibid.). However, some writers feel the focus should also be on reducing the river's sediment load, arguing that a failure to do so will mean there will not be enough flow in the main channel to flush the particles out to sea. Sediment accumulation in the main channel will increase, further worsening the flood discharge capability of the river and possibly leading to future disasters.

The changing pattern of the water budget for the Yellow River basin can be seen in the following data. On average the river receives an input of surface water from all tributaries in the basin at an estimated rate of 58 billion cu. m annually. Of this, in 1980 27.1 billion cu. m was diverted for irrigation and other uses, but with water consumption increasing at the rate of 4.5% each year, by 1987 water diverted from the river shot up to 37.6 billion cu. m. Underground-water extraction also increased from 6.4 billion cu. m to 15.9 billion cu. m. during the same time period. The estimated underground-water reserves of 19 billion cu. m are being rapidly depleted.

All this time, about 32 billion cu. m of water flowed out to sea, taking with it the river sediment. The average annual sediment load is estimated to be 1.35 billion tons, which needs a minimum of 20–24 billion cu. m of the annual flow to achieve an efficient flushing process. Hence, so the theory goes, if the sediment load in the river could be reduced by half, 10 billion cu. m of water presently wasted on sediment flushing could be put to better use and the need for expensive diversion from southern China be considerably reduced.

A member of the Institute of Geography of the CAS, Jia Shao-feng, proposed such an idea in an article in the obscure journal *Ke Ji Dao Bao* (issue No. 9, 1994) entitled 'A substitute proposal for the North–South water diversion – water and soil conservation in the middle Yellow River valley'. An abridged version appeared in the journal *People's Yellow River* the following year (No. 3, 1995). In essence, his argument was that the problems of the Yellow River could never be fundamentally solved without reducing the sediment content by controlling erosion in the loess region. Human efforts to restrain nature's plain-building process by channelling river flows with levees or de-silting reservoirs could only provide temporary and marginal solutions; the key effort must rest with reducing erosion and sediment. According to Jia, it was generally estimated that the area of high sediment yield in the middle Yellow River basin, with an annual erosion rate exceeding 1000 tons per sq. km, extended for 430,000 sq. km over a landscape sparsely populated and totally dissected by gullies (mature gullies reaching heights of 200 to 300 metres). Due to the enormous scope for comprehensive erosion control and little immediate economic return to pay for the expenditure, erosion control and sediment control were not being treated as an engineering project to be completed with targeted objectives and dates. Rather, they were left as incidental tasks to be accomplished whenever or wherever they could produce agricultural benefits. His article argued that it was feasible to achieve comprehensive sediment control for the entire middle Yellow River valley by treating it like an engineering project with modern construction methods, to be completed within, say, 20 years at a cost comparable to a single hydro-electric project.

In 1955, a department under the Yellow River Conservancy Commission, the Upper and Middle Yellow River Administrative Bureau was established, and given the responsibility and resources to initiate soil conservancy and sediment reduction projects across the loess region. Ever since the early 1950s, hillside terracing has been greatly promoted as it created arable land and at the same time reduced erosion damage. The drawback to terracing is its labour-intensive nature and the relatively low productivity of the terraced plots. As most of the loess highland is only thinly populated, it is not likely that terracing can be an effective means to implement comprehensive erosion and sediment control. Even in densely populated areas it is becoming increasingly difficult to mobilise peasants to contribute their labour, with minimal compensation, to terrace the gully slopes, so that additional acreage from terracing has been slow to appear in recent years.

Development of grassland is most successful in the relatively flat terrain where ground water is plentiful, as in the regions around the northern bend of the Yellow River in Inner Mongolia, where natural grassland already exists. Large-scale grass or shrub planting and reforestation in the badly dissected loess plateau has achieved only limited success due to the aridity of the region; the reported survival rate of new trees is only 20 per cent. Damage done to the grassland by goat herding and reckless destruction of young trees for fuel has also limited the the effort's success. Furthermore, over terrain dissected almost totally by gullies with steep slopes and poor water retention, grass, shrubs, and trees have difficulty taking roots. However, erosion is actually the most severe on such terrain, and thus these plantings alone, without subordinate measures, would not be effective in achieving comprehensive erosion control.

An approach which is considered more promising is to first improve the productivity of the land with sediment plots by damming gullies with 2–5 m-high earth dams, called silt-trapping dams (or warping dams), trapping sediment from eroded topsoil behind them and forming high-yield plots within a few years. These dams can be built at a relatively low cost, one upstream from the other in a staging manner, and the peasants much prefer them to the terraced plots. Driven by economic incentives, local initiatives are propagating the sediment plots with only limited assistance by the government. Their development contributes substantially to reductions in sediment entering the rivers.

To fully develop the benefits of the sediment plots, the concept of a core project had been advocated ever since the 1980s. A core project applies to a major gully with a watershed of 3–5 sq. km, which will be provided with a 30 to 50 m-high key dam. Its reservoir, with an average capacity of one million cu. m, is adequate to withstand exceptional storm runoffs of the 200-year level even after 15 years of sediment accumulation. Hillsides surrounding the reservoir will be terraced, creating irrigated high-yield plots, which in turn will help to reduce slope erosion and silting of the reservoir. Sediment plots created downstream from the dam will be protected by the dam from severe runoffs, and those created upstream from the dam in the branch gullies will trap sediment from reaching the dam reservoir. This concept of 'agricultural oases' is being recognised as a promising development. By promoting the extensive construction of such projects in areas with erosion rate exceeding 5000 tons per sq km, it is being claimed that sediment entering the rivers downstream can be greatly reduced.

But, while the construction of core projects based on local initiative provides a viable means of reducing sediment in the rivers, its implementation relies heavily on the demographics and economic development of the region. With 80% of the heavily dissected loess plateau unpopulated, or thinly populated, it is unlikely that sediment in the Yellow River can be much reduced by such regional efforts in the near future. A major influx of funds from the central government for the region's reclamation to promote economic development and population settlement is needed for achieving the final sediment control in the Yellow River. And, as has so far been demonstrated, the government is focussing more on increasing the water flow via the link-up with the Yangtze.

At the same time, the government is intent on accelerating the development of water resources on the upper reaches of the Yangtze as part of the programme to develop the entire western part of the country. Liu Junfeng, general manager of the Ertan Hydroelectric Development Company, for instance, revealed that four large hydropower stations with a combined installed capacity of 9.05 million kW would be built on the Yalongjiang River in Sichuan Province, which is one of the country's 12 large water energy resource bases (proven reserves of water energy: 33.4m kW). In fact, up to 21 hydropower stations could be built on the river in accordance with future development plans (*Xinhua*, 28 February 2000).

Of the four initially planned, the largest is the Jinping No. 1 hydropower station, with an installed capacity of 3.6m kW, capable of generating 18.2

billion kWh a year, whose construction will cost an estimated 30 billion yuan. The Ertan Hydropower Station, the largest facility of its kind in China at the time of writing, is also located on the Yalongjiang River. Its installed capacity of 3.3 million kW.

Technical concerns

Returning to the proposed Yangtze–Yellow River diversion plan, we need finally to consider technical considerations that could make such a re-engineering of nature unwise. As discussed earlier, the basic assumption in all the diversion schemes is that the Yangtze and its tributaries have more than enough water to spare, and that that there is no fear of them drying up in the foreseeable future. But can this be taken for granted?

One man who does not think so is Professor Lei Hengshun of the Yangtze Technology and Economy Society, who asserts that the Yangtze will become just as dry as the Yellow River by 2020. In 1997, the Yangtze had 15 per cent less water than normal, according to the Wuhan Shipping Department. 'This is a dangerous signal. If the trend continues, it could bring incalculable losses and irreversible damage to the ecology of the vast basin area' (*China Economic Times*, 6 April 1998)

The professor said the Yangtze basin's ecology was in dire crisis, not only because of increasing soil erosion and silting up, but through contamination. 'There is nearly no clean water in it from beginning to end', he said. 'One day soon, those living on its banks will have to search hard for drinkable water.'

Chongqing, whose dirty reputation was discussed in Chapter 5, accounted for more than 70 per cent of the river's pollution. Only 40 per cent of its industrial waste water met national standards and nearly all its sewage was still being discharged directly into the river without treatment in 1997, when the city began building its first water disposal plant, Professor Lei claimed. Further downstream, he said, the situation was just as bad, with cities like Wuhan, Nanjing and Shanghai all adding their share of filth. The Yangtze was also being filled with garbage and sewage from shipping. Each boat passing through the Three Gorges generated between 177–236 tonnes of waste a year and faeces from 10,000 passengers were discharged each day. 'This is far beyond the river's natural ability to cleanse itself', he insisted (ibid.).

In addition, grandiose civil engineering schemes to improve the Yellow River have so far not always enjoyed great success. In a precursor to the present arguments raging over the mammoth Three Gorges Dam, we have the example of the similarly named Sanmenxia (Three Gate Gorge) Dam, built on the upper reaches of the Yellow River at the height of the Maoist love affair with 'bigger is better'. The author Lynn Pan, who journeyed along the Yellow River in the 1980s, visited the dam site and noted:

> The dam structure was completed as early as 1960 . . . but very soon afterwards the river's silt clogged up the reservoir, and the Chinese engineers

found they had to rebuild the dam and overhaul their plans for the auxiliary works. The power plant's kilowattage was reduced from a million to a measly 200,000, and though the reservoir's storage capacity runs to billions of cubic feet, there is no way that all of it can ever be used. So what was to be a showcase and fulcrum of Yellow River control now stands as a monument to hasty miscalculation and adventure.

(Pan, 1985, p. 49)

According to Pan, the Chinese planners had been too sanguine to be seriously bothered about the silt problem when the project was still in its earliest stage, and had plunged into its construction heedless of the suggestion that silt discharge gates be built at the base of the dam, or that the sediment be collected at one or more dams upstream before they started on the one at Three Gate Gorge.

Such was the boldness of the conception that, of the six designs submitted for the main dam, the one selected was the most expensive. As so often happens, the work came to cost more and entail greater population relocation than had been originally estimated. The initial investment had to be doubled and two and a half times the number of inhabitants moved than had been anticipated. It was an expensive business from the start, for most of the equipment, the heavy earth-moving machinery, the turbines and the like, had to be bought from the Soviet Union, Eastern Europe and Japan. Yet in the end this expensive marvel had to be whittled down to size, and when all the dust had settled, and the project put into operation at last, its effectiveness was no more than that of the cheapest of the six designs considered.

(Ibid, p. 50)

The Yellow River's famous sludge and its potential for havoc had yet to be grasped. The engineers had thought that not for another fifty years would the reservoir's capacity be significantly reduced by silt, but within five years of the dam's completion not only the reservoir but also the valley all the way to Xian was threatened. The reconstruction then began in earnest. Tunnels were dug; pipes were pressed into service as silt-evacuation conduits; dynamos resistant to silt abrasion were installed; and the silt, as it entered the reservoir, was sluiced out of the dam through eight openings originally intended for water diversion during the actual construction.

Yet, despite American and Japanese warnings, the site was too inviting to be ignored. Set against the geological fact that this was the only stretch of river in the vicinity not susceptible to earthquakes, the disadvantages did not seem quite so overwhelming; set against the hydrographical fact that more than 90 per cent of the basin's drainage area and most of the main tributaries lay upstream from this point, making it a veritable fulcrum of flood control, the disadvantages seemed hardly worth considering. These same arguments are will undoubtedly come into play with the north–south water diversion project.

But apart from any technical objections, is the water diversion scheme necessary? Geng Shufang, a research fellow with the Institute of Geology of the China Academy of Geological Sciences thinks not. He argues that it is a waste of time and money, and that a lower level of investment would yield quicker results if directed instead to more efficient use of existing water resources of north-west China to meet local needs.

He admits the region has always had less water than the lush area of the south, but acute shortages had never occurred to threaten local economic and social development in the past. While the region certainly lacks surface water, he argues there is no shortage of ground water. Geng has created a detailed blueprint for tapping melted glacier water and tapping underground reservoirs hidden by the shifting sands of the vast deserts now covering the far west, schemes that he insists will meet all the needs highlighted by the government's Yangtze diversion plan (author interview).

Zhang Jiacheng, Vice-President of the China Institute of Meteorology, meanwhile, is advocating more efficient agricultural and industrial methods needing less water, which can be obtained from rainfall and desalinisation of seawater (author interview).

But both these alternative proposals seem unlikely to be heeded. The government, it seems, is now determined to fulfil Mao's dream.

8 Ecological tramplings

For some years now, the concept of 'ecological footprints' has gained currency in the study of the impact of development, especially urban development, on the ecosystem. Unfortunately, in many if not most cases the impact produces more than 'footprints'. Therefore we have elsewhere (Cook and Murray, 2001) coined the term 'ecological tramplings' to highlight the negative impact of urban and industrial growth especially on surrounding areas. China's economic expansion of recent decades has been tremendous; the concomitant urbanisation associated with this has also been on a massive scale. So too has been that in 'Greater China', including Hong Kong and Taiwan (see Chapter 9).

In Chapter 5, we considered the impact of growing urbanisation. Here, we simply summarise the negative ecological impact of this rapid urbanisation and its interrelated industrialisation:

- **Land** lost to agriculture at a rapid rate, along with a decline in soil quality due to pollution
- **Air** quality declining due to vehicle and industrial pollution
- **Water** quality declining for the same reasons; in addition, water shortages are increasing because urban populations and industries are high users of water
- **Resources** depleted to feed urban growth
- **People** are contributing to, and being affected by, modernisation, leaving rural areas for cities and having high material expectations which impact directly or indirectly on the environment
- **Buildings** become taller, increasing the use of energy and resources compared to traditional urban forms.

In contrast, positive ecological features hinge on:

- **Technological ingenuity** that can focus on such issues to find novel solutions to them
- **Environmental awareness** to ensure that solutions are prioritised, regulations developed and enforced at the national not just the local level.

Atmospheric pollution

Atmospheric quality across China is such that residents and visitors alike cannot fail to notice the high levels of atmospheric pollution that are endemic to many towns and cities. This realisation is being increasingly backed up by official data gathered by municipalities, government agencies and external bodies such as the World Health Organisation (WHO). Writing at the end of 1998, Chen Qiuping notes that:

> Five of the ten most seriously polluted cities in the world are located in China. Urban residents are breathing foul air where the average daily value of total suspended particulate is ten times higher than the international standard. According to related data, the death rate from lung cancer in urban areas is increasing annually.
>
> (Chen Quiping, 1998, p. 8)

WHO data suggests that China's capital city, Beijing, which should be at the forefront of pollution control, was instead the third most polluted city in the world in 1998, a damning indictment of the past lack of concern for environmental matters not just in the city itself, but also nationally at government level. Nationally, in the two decades to 1990, China's sulphur dioxide emissions grew by more than a factor of three, and by 2020 may have increased by a further factor of three (Foell *et al.*, 1995) due to this 'business-as-usual' attitude. Atmospheric pollution has a range of causes, and includes dust from the many construction sites, discharges of sulphur dioxide, nitrogen oxide and total suspended particulates (TSPs) from coal burning, nitrogen oxide, lead and other emissions from vehicle exhausts plus a range of discharges from industrial plants.

As regards the latter, in 1999 it became apparent that the Shougang steel plant was now a major target for pollution control, particularly in the light of Beijing's bid for the 2008 Olympics (the city failed by only two votes to obtain the 2000 Olympics, rather than Sydney). During the visit of the head of the International Olympic Movement, Juan Antonio Samaranch, prior to the vote last time, it transpires that the Shougang plant was required to close for the day to reduce pollution levels in the city.

O'Neill (2000) notes that in 1999, pollution levels in the district around Shougang were above the permitted levels 85 per cent of the time, and the pollution was spreading across the city due to the winds from the north-west, leading delegates from Beijing to the highest law-making body, the National People's Congress, to file a protest against the level of pollution from the steel plant, saying it was hindering the city's progress and its bid to hold the Olympics. Despite its 7 million tons annual output it is possible that the plant, which does seem to us to be an anomaly for a nation's capital, especially an inland one, may well be relocated, perhaps to Rizhao, a city of 200,000 on the east coast of Shandong Province, that has good transport facilities by road, rail and sea. Rizhao has spare land, a deepwater port that handles between 17 and 18 million tonnes of cargo a year and the capacity to double that, one large and two small reservoirs, a surplus of electricity, two railway lines and three major roads, and was a strong competitor for the plant, with Japanese and German equipment, that eventually went to Baoshan, north of Shanghai, in the early 1980s.

The provincial Government of Shandong has established 'a big steel plant' office to lobby the Capital Steel Plant (Shougang) and argue the case for a move to Rizhao as opposed to sites in neighbouring Hebei province, such as Tangshan and Handan, which are also trying to lure it (ibid.). O'Neill notes unofficial speculation that the Shijingshan site could become 'a Disney park' and that 'Surely Mao's reaction to that would be unprintable' (ibid.)! The move of this major plant would permit further modernisation, and also feed into the policy of closures of inefficient, high polluting and obsolescent small plants across China, leading to environmental improvements and longer-term economic gains but with a potential social cost as redundancies ensue (*Business Weekly*, 9 January 2000).

Coal burning is another serious issue (just how serious, in terms of human health, is the topic of an ongoing study, summarized in Box 8.1). In Beijing, for example, coal burning contributes 90 per cent of sulphur dioxide, 50 per cent of nitrogen oxide and 40 per cent of TSPs in the atmosphere (in addition, construction sites are responsible for 50 per cent of dust, and nitrogen oxide from vehicle emissions 'is reportedly the number one winter pollutant in Beijing') (Xiao Chen, 1999, p. 14). The city, like many across north China, burns a huge amount of coal in the winter months, with 8 million tons used for heating in 1998–9 during the four-month heating period, November–February (ibid.). However, the city's environmental department insists that things are improving, with the reduction of coal use and adoption of alternative energy sources (see Table 8.1).

China as a whole, however, remains heavily reliant on coal, often of a high sulphur content. By 1999, across China:

> more than 23 million tons of sulphur dioxide are discharged into the atmosphere every year, causing acid rain to fall on 40 per cent of the nation's lands, according to the environmental administration.
>
> (Guo Aibing, 1999, p. 2)

Further,

> acid rain has affected almost all areas south of the Yellow River except the Qinghai–Tibet Plateau. Professor Wang Wenxing from the Chinese Academy of Sciences has estimated that the annual losses from acid rain in China top 30 billion yuan.
>
> (Chen Qiuping, 1998, op. cit., p. 9).

Table 8.1 Beijing's achievements and goals

Sector	Current achievements	Future goals
Energy	Coal use down to 50%, compared with national average of 75% Switch to LNG for most households	Reduction to 20% through more energy-efficiency savings and fuel substitution
Vehicle emissions	CO reduced by 18% 1998–2000; NO down 17%	
Public transport	60% of buses run on CNG 50% of taxis use LPG	Industrial emissions reduced by 30% by 2005
Industry	Chemical, cement and metal producing factories closed down or switched to new, non-polluting products.	

Source: Data provided by Beijing Environmental Protection Bureau.

The Chinese Research Academy of Environmental Sciences estimated that 40 per cent of the country was affected by acid rain, causing 1.6 billion US$ worth of damage to crops, forests and property annually in the mid 1990s (Walsh, 1995). And the transport of sulphur to other parts of Asia, as we discussed in the preface is becoming an area of increasing international environmental interest and concern (cf. Carmichael and Arndt, 1995, and Robertson *et al.*, 1996, Arndt *et al.*, 1996, Sato *et al.*, 1996, and Sharma *et al.*, 1995, all cited in Carmichael and Arndt, 1998). The present situation in northeast Asia is that high levels of acidic compounds, predominately sulphate and nitrates, are being deposited throughout the region. The levels of acid deposition are sufficiently high to put ecosystems at risk (as estimated by critical loads for Asia) in vast regions including southern China, South Korea, Taiwan, and southern Japan. The situation would be worse if it were not for the fact that north-east Asia has high levels of wind blown soils and high levels of ammonia that neutralize an appreciable fraction of the strong acids. Model calculations suggest that trans-boundary transport already contributes to acid deposition in Korea and Japan (Carmichael and Arndt, 1998 op. cit.).

Present trends in energy consumption throughout Asia, with China at its heart, impose significant environmental threats to a variety of ecosystems in large parts of the region. Carmichael and Arndt suggest that over the next decades, as the regional sulphur dioxide emissions increase, sulphur deposition levels will be higher than those observed in Europe and North America during the 1970s and 1980s, and in some cases will most probably exceed those observed previously in the most polluted areas in central and eastern Europe. This increase in sulphur dioxide emissions will severely threaten the sustainable basis of many natural and agricultural ecosystems in the region. Taking the critical loads as an indicator of sustainable levels of acid deposition, future sulphur deposition will exceed critical loads by more than a factor of ten in wide parts of Asia. These levels of sulphur deposition would cause significant changes in the soil chemistry over wide areas in Asia, affecting growing conditions for many natural ecosystems and agricultural crops (Carmichael and Arndt, op. cit.).

Acid rain corrodes metal, often 'eating' lampposts and buses for instance, attacks buildings and trees, and leaches through soil into watercourses, thus endangering fish species. If produced by emissions from coal-fired power stations it can be treated, and minimised, at source, but for general industrial or domestic use the only solution is to use low-sulphur coal or no coal at all. The latter is just not feasible for the PRC, which produced 1.25 billion tons of coal in 1998, albeit that this was a decline of 8.9 per cent from 1997 (China Statistical Information and Consultancy Centre, 1999, p. 11). 'In the absence of an unforeseen technological breakthrough, coal will continue to be China's major source of energy in the medium- to long-term future . . . coal production is expected to increase by 40 million tons per year to around 2070 million tons in 2010' (Ögütçü, 1999, p. 93).

Consequently, the State Development Planning Commission has been charged to oversee the development of clean coal technology, including utilisation of high-quality, low-cost coal reserves in West China (ibid.). Low-sulphur coal is being promoted in Beijing for instance, where the Beijing Municipal Environmental Protection Bureau has introduced regulations and a system of spot-checks on businesses and 8,000 yuan fines for winter season 1999–2000 to ensure that ordinary unwashed coal is not used (Tang Min, 1999, p. 16).

What are the alternatives to coal? Oil exploration, especially in the far west and offshore, notably in the South China Sea, is being heavily promoted, but supply has struggled to cope with growing demand. China began importing oil in 1993 and within two years was purchasing 400 thousand barrels per day on international markets. This doubled to 800 thousand barrels per day in 1997 and is projected to reach 7–8 million barrels per day in 2015 and 13–15 million barrels per day by 2025 (Takle, 1998). This makes it difficult for oil to compete with domestic coal as an energy source.

Nuclear power will be one of the preferred alternatives; hydro-electric power the other. The latter has been discussed indirectly in Chapter 6 via the Sanxia Dam controversy, for which provision of hydro-electric power is one of its main objectives, and has been shown to have a number of problems attached to its development. Nuclear power currently contributes a small fraction of energy supply – one per cent in 1999, hopefully rising to three per cent in 2000 and five per cent in 2020 (*Agence France Presse*, 2 June 2000).

'China's nuclear capacity is currently 2.3 GW. Tending toward the conservative, we assume that nuclear capacity could reach 10.7 GW by 2010' (Ögütçü, 1999, p. 103). The disaster of 1986 at Chernobyl in the Ukraine, however, showed that the concerns of the anti-nuclear lobby were not

Box 8.1 'Coal fumes poison millions'

Millions of Chinese are being poisoned by fumes from the raw coal they burn in their homes for cooking and heating, and people in other developing nations may face similar risks, according to a study released in March 1999 by a group of American and Chinese scientists. An estimated 800 million Chinese use coal in their homes. But in many rural communities, the fuel is full of arsenic, mercury, fluorine, lead, and other poisonous metals that can pose a serious health threat, the study reported. The poisons become part of the smoke from burning coal and are then breathed into lungs and baked into foods, according to one of the report's authors, Harvey E. Belkin. For example, in Guizhou province, which has beds of arsenic-rich coal, farmers routinely dry peppers over coal-fired stoves. The produce absorbs the smoke and fumes – imparting up to 500 parts per million of arsenic to each pepper, a potentially dangerous level – and then are added to virtually every food prepared. It was not known whether the arsenic poisoning has caused any deaths, Belkin said, because death certificates and autopsies were rare in rural China, but at least 3000 cases of chronic arsenic poisoning had been confirmed, with symptoms including skin cancer and open sores. 'The coal we saw in China could never be burned in the United States' because of the poison mineral content', said Belkin, adding, that there was 'high arsenic, high selenium and high mercury [levels]'. Some of the arsenic-rich coal samples also had high concentrations of antimony (up to 200 ppm), gold (almost 600 ppb), and mercury (up to 45 ppm), the study reported.

Another of the report's authors, Robert E.Finkelman, said that, in another area studied, farmers used coal with a high fluorine content to dry corn. The grain takes up high levels of the poison mineral, and as a result fluorosis, or fluorine poisoning, was very common an estimated 10 million cases. The effects of this can range from discoloured teeth to softened, twisted and crippled bones. About 22 per cent of rural homes depend entirely on coal. 'If they want to cook or heat, they have to use coal', Belkin explained. For many peasants, the coal is dug out of hillsides, free for the taking. Many poor Chinese, however, burn this coal in unvented stoves, filling their homes with fumes. Indeed, most homes in rural villages are wreathed with the bluish haze of coal smoke in the early morning. (Belkin *et al.*, 2000)

misplaced, especially in a political system with a high degree of closure and secrecy rather than openness, and worries remain concerning nuclear waste disposal and a potential 'China crisis' in its nuclear power stations, especially when these are being built within the highly populated east coast region. As with acid rain, radio-active clouds, too, have a long reach and could contribute to potential conflict situations in Pacific Asia. Nevertheless, the Chinese Government is proceeding slowly with the construction of new nuclear power stations, and is building a small, experimental fast-breeder reactor at Fangshan, 50 kms south-west of Beijing, with Russian help. The Russians are also expected to be involved in the construction of two uranium enrichment plants in the northern cities of Lanzhou and Hanzhong.

China has substantial hydropower potential, much of which remains untapped. Solar and wind energy have not been exploited. The biggest obstacle stopping full utilization of solar power now seems to be lack of government involvement. 'Technology is no longer a problem', insists Zheng Ruicheng, senior engineer with the Chinese Academy of Building Sciences. 'Most technologies in this field are mature.' But the cost, about five to eight times higher than regular energy such as thermal power and hydropower, has frightened away many potential investors, according to Zhao Yuwen, vice-president of Chinese Solar Energy Society, adding that, without preferential policies, it would be almost impossible for the expensive photovoltaic electricity to compete with other energy sources. In addition, China had no laws or regulations boosting development of renewable energy, he said (author interviews).

There is some business interest in developing solar power. Changzhou Trina Aluminium Panel Curtain Wall Manufacture Co. Ltd in Jiangsu Province, for example, was planning to build a solar-powered gymnasium in Beijing to support the capital's hosting of the 2008 Olympics, and also promote the technology. 'The market potential is huge', insisted Gao Jifang, the company's general manager. 'There are at least 14 million families and 60 million farmers who have no access to electricity, but live in places of rich solar resources. They are potential customers.'

Biomass has seen limited use for energy, although multiple (double and in some cases triple) cropping patterns, with resulting requirement to remove dead plant material for replanting, would seem to offer potential for future growth of biomass for fuel. Drennen and Erickson (1998) assert that, even under optimistic scenarios of foreign investment in hydro, nuclear, and biomass, the primary energy supply in China in 2025 will be fossil fuels (68% coal and 25% oil). In Box 8.2, we give further consideration to coal's future (Tables 8.2 to 8.5, meanwhile, provide data on the growth and composition of China's electric power system). This means that the acid rain problem will continue, affecting Japan and other countries due to the spatial extent of its impact, contributing to tensions and pressures for alternatives to coal use (Cook and Murray, 2000, op. cit., Ögütçü, 1999, op. cit.).

Better energy efficiency

As has been discussed in earlier chapters, China's energy efficiency is inadequate, with a high-energy consumption per unit commodity pro-

Table 8.2 Electric power development 1988–99

Year	Installed capacity (GW)	Growth rate (%)	Power generation (TWh)	Growth rate (%)
1988	116	12.2	545	9.6
1990	138	8.9	621	6.3
1993	183	9.8	836	10.9
1995	217	8.7	1,007	8.5
1996	232	6.9	1,075	6.8
1997	254	9.5	1,134	5.5
1998	277	7.9	1,158	2.1
1999	294	7.3	1,230	6.2
Average 1988–99		9.1		8.3

Sources: *China Energy Annual Review 1997, China Statistical Year Book 1999, Financial Times,* 17 January 2000 in Zhou, 2000 (op. cit.). Excludes nuclear, cf. Table 8.3.

Table 8.3 Installed capacity and power generation 1997

	Thermal	Hydro	Nuclear	Total
Installed Capacity (GW)	194	80	2	256
Power generation (TWh)	929	195	14	1,148

Source: *China Electric Power Statistical Yearbook 1997* in Zhou, 2000 (op. cit.).

Table 8.4 Power consumption by sector 1987–97

	1987	1990	1995	1997
Total Consumption (TWh)	490	613	1,002	1,104
Share by sector (%)				
Industry	81	79	74	73
Heavy industry	65	63	59	58
Light industry	17	16	15	15
Agriculture	7	7	6	6
Commercial	5	5	6	8
Residential	6	8	10	11
Other	3	3	3	3

Note: Rounding off may result in totals other than 100 per cent

Source: *China Electric Power Statistical Yearbooks 1991–98* in Zhou, 2000 (op. cit.).

duction, while the country's energy pattern focusing on coal has had enormous impact on environmental pollution. This, however, can be reduced substantially through energy conservation, and the development of new energy and renewable resources. So, what are the prospects? The following section is based on research carried out by one of the authors for a major Japanese company in 2001, and is drawn from interviews with government, industry leaders and various analysts.

Box 8.2 Future options of King Coal

A report published in early 2000 by the US-based Pew Centre on Global Climate Change (set up in 1998 by the Pew Charitable Trusts to promote a new cooperative approach and critical scientific, economic and technological expertise to the climate debate) argued that in the next 15 years the principal drivers of the technology choices for the Chinese power industry would be:

1   Growing awareness that under a business-as-usual path, carbon emissions from thermal plants will increase from 189 million tons in 1995 to 491 million tons in 2015, and sulphur dioxide emissions from 0.5 to 21 million due to heavy reliance on coal-fired power stations.
2   Increasing demand-side energy efficiency by 10 per cent from business-as-usual projections could reduce carbon dioxide and sulphur dioxide emissions by 19 and 13 per cent in 2015.
3   Expanding the availability of low-cost natural gas through market reforms (also discussed in the main text) could reduce emissions of the two gases by 14 to 35 per cent.
4   Accelerating the penetration of cleaner coal technologies could help China reduce sulphur dioxide and particulate emissions, but the associated impact on carbon emissions would be minimal.

Faced with projections of a three-fold expansion in power demand by 2015, and with coal remaining the largest and lowest-priced energy source, the report said that, disregarding health and environment costs, China's power generation would become even more intensive, with attendant environmental costs. As this was virtually unthinkable, the report, prepared primarily by the Beijing Energy Efficiency Centre and China Energy Research Institute, put forward six considerations for a more sustainable future:

1   China needs to continue to develop and deploy technologies that lower sulphur emissions (e.g. coal washing, flue gas desulphurisation, clean coal combustion technologies).
2   Natural gas can help reduce sulphur dioxide, carbon dioxide and other pollution from power generation, but gas availability and price will critical.
3   As nuclear power is not economically competitive with other options, it can play a larger role if China chooses it to lower polluting emissions or for energy security.

4   The country's remaining hydro-electric resources are located in remote areas, making development expensive, but efforts should continue to be made to change this.

5   Renewable energy alternatives like biomass, wind energy and geothermal cannot yet compete, but China should continue to work on cutting costs to change this.

6   China must continue to promote energy efficiency (not a strong point in the past) as part of raising the competitiveness of consuming industries (Zhou *et al.*, 2000).

Table 8.5 Regional power consumption and growth rates

	1997 (TWh)	Annual growth 1990–97 (%)
North	335	4.8
East	269	7.7
Central	159	7.7
Guangdong	79	8.2
South-west	117	10.7
National total	1,104	8.7

Sources: *Study on Alternative Energy and Energy Supply Strategies in China*, 1998; *International Energy Outlook*, 1998, cited in Zhou, 2000 (op. cit.).

According to the state energy conservation plan, the long-range energy conservation target is equivalent to 70 million tons of standard coal each year. This is considered a vital component of the current industrial restructuring plan, whereby the government is trying to haul the most vital state-owned enterprises out of the cycle of chronic debt, production inefficiency and spiralling costs.

Government experts describe the key issues as follows.

First, increasing investment in the energy conservation and new energy is vital to promoting economic development (see the next section). For the state, carrying out construction through investment will help key energy-consuming industries to adjust their structure, to quicken the pace of pulling large and medium-sized state-owned enterprises out of difficulty and to enhance national strength. For enterprises, this will help to achieve optimisation of their product mix, lower energy consumption and production costs, improve their product quality and increase their market competitiveness. For society, this will not only help to improve the environmental quality, but also will be of great pragmatic significance to expanding domestic demand, stimulating market demand, improving productivity, increasing job opportunities and ensuring social stability.

Second, expanding energy conservation and investment in new energy are feasible. At present, most state-owned enterprises have a high asset debt level and insufficient funds of their own, making banks reluctant to extend loans. As a result, the enterprises cannot undertake energy-efficient environmental protection projects that require a small amount of investment and can yield quick returns. Therefore, the state should establish energy conservation and environmental protection industrial funds by subsidizing project capital and through the issue of construction bonds, in order to provide policy-oriented assistance. The move can greatly increase banks' confidence in extending loans and arouse the investment initiative of localities and enterprises, and effectively spur loans from commercial banks and fund inputs of localities and enterprises.

Government experts estimate that if the country invests 20 to 30 billion yuan in the energy conservation and new energy area each year, with the state contributing one third and the rest raised by enterprises and from bank loans, a new energy conservation capacity of seven to ten million tons of standard coal will be added. The discharge of carbon dioxide will be cut by 18.6 to 26.6 million tons annually, the emission of sulphur dioxide will be reduced by 270,000 to 380,000 tons a year, the discharge of dust will be cut by 100,000 to 150,000 tons annually, the discharge of slag will be reduced by 1.8 to 2.6 million tons a year. The GDP will be increased by 26 to 40 billion yuan annually.

The state began to extend special energy-conservation loans for capital construction in 1981. Since then, the China Energy Conservation Investment Company (including its predecessor, the Energy Conservation Bureau of the State Planning Commission and the Energy Conservation Company of the State Energy Investment Corporation) has been in charge of 'special energy conservation capital construction projects'.

By the end of 1998, the state special energy-conservation plan for capital construction had handled 2,280 projects with a total investment of 61.5 billion yuan, covering areas such as integrated production of heat and electricity, central heating supplies, the utilization of exhaust heat, wind power generation, the revamping of rural power grids, building materials, metallurgy, non-ferrous metals, chemicals, petrochemicals, machinery, the light industry, electronics, textiles and coal. The state contributed six billion yuan, commercial banks loaned 20 billion yuan and local governments and enterprises were 'guided' to invest 35.5 billion yuan.

The SECIC calculates that these projects have formed an annual energy conservation capacity of 45.8 million tons of standard coal. They have increased the GDP by 80 billion yuan a year, cut the discharge of carbon dioxide by 121.7 million tons a year, the emission of sulphur dioxide by 1.74 million tons annually, the discharge of dust by 700,000 tons a year and the discharge of slag by 12 million tons annually.

An analysis shown to Murray by a SECIC official cited the example of the Huaihe River Basin, where projects such as those for the recovery of soda ash from the paper-making process and alcohol liquid had, since 1996, enabled the State to meet its environmental protection goal of treating 110 million tons of sewage a year and reducing the discharge of chemical oxygen demand by 300,000 tons and biochemical oxygen demand by 100,000 tons annually, thus making great contributions to the pollution control in the basin.

Marked progress has been made in the undertaking of new energy projects in recent years, according to the analysis. Wind power generating units with a total capacity of ten megawatts have been installed in Nan'ao, Guangdong Province, Huitengxile, Inner Mongolia and Dongfang County, Hainan Province. Solar energy power plants with a combined generating capacity of 55 kW have been installed in three counties in Tibet that used to have no access to electricity, including Shuanghu and Cuole.

A number of projects for comprehensive utilization of terrestrial heat have been completed in Zhangzhou, Fujian Province. As a result, terrestrial heat resources have been used for industrial drying, regional heat supply and fish breeding, with high economic returns and good social effects.

Energy conservation

The energy conservation efficiency of integrated production of heat and electricity is now acknowledged throughout the world. Generally, heat efficiency of thermal power plants is 36 per cent and that of integrated production of heat and electricity is 60 per cent. If power output from integrated production of heat and electricity is calculated at 100 billion kWh and heat supply at 950.67 million joules in 1997, the amount of energy saved would come to 21 million tons of standard coal, the emission of carbon dioxide would be cut by 55.8 million tons, that of dust would be reduced by 290,000 tons and that of sulphur dioxide would be cut by 800,000 tons a year.

China's current heat and electricity generating capacity is 22.22 million kW, 12.12 per cent of its thermal power generating capacity. The use of central heating in place of scattered low-efficiency boilers has shown good results in energy conservation and environmental protection. However, central heating systems currently cover only an estimated 800 million sq. m (with integrated production of heat and electricity covering 500 million sq. m) of buildings, with the heat conversion rate standing at only 12.24 per cent. In particular, the northern cities with an annual heating period of 3000 to 4000 hours still have a total floor space of nearly 1.6 billion sq. m relying on small boilers for heating.

Therefore, major industrial cities in northern China need to build heat and power plants with a generating capacity of 100 to 200 megawatts to form a large heat supply network for eight million sq. m. Large numbers of small and medium-sized cities need to build small and medium-sized heat and power plants to form a heat supply network for one to two million sq. m. If the heat conversion rate in northern China increases from 29.8 per cent to 50 per cent, the floor space with access to central heating will increase by 500 million sq. m. If all heat supply comes from integrated production of heat and electricity, investment in heat networks will amount to 25 to 40 billion yuan and construction of heat sources will need investment of 90 to 130 billion yuan.

Key energy conservation investment areas from now are:

- Integrated production of heat and electricity and integrated production of heat, electricity and gas, or heat, electricity, and cool air)
- Improved central heating and the revamping of industrial boilers
- Upgrading of industrial kilns and furnaces and the utilization of exhaust heat
- Development of urban gas networks and the recovery of wasted gas
- Shaped coal and coal with a fixed percentage of components
- Energy-efficient buildings
- Revamping of metallurgical kilns and furnaces and pressure differential power generation
- Coal washing and dressing and comprehensive utilization
- Upgrading of chemical industrial technology
- Revamping of oil refining technology
- Upgrading of the light industrial technology and environmental protection projects
- Revamping of building material technology and new wall materials
- New machinery and electrical products and energy-efficient meters and instruments
- Major general-purpose equipment (such as the electric motor traction system, boilers, non-crystal metallic transformers, highly efficient electro-optical sources and valves)
- Wind power generation, solar energy generation and new energy projects.

New and renewable energy

China abounds in new energy and renewable resources. But hydropower resources developed so far account for only 11 per cent of the total of 378 million kW, while the country's wind power resources stand at 1.6 billion kW, 15 per cent of which can be developed and utilized with available technology. On two thirds of China's landmass, the annual solar energy radiation exceeds 600,000 joules/per sq. cm, with vast development and utilization prospects. Yet such resources are being wasted because of the low development and utilization level. By 2000, for example, the government reported a national wind power generating capacity of 400 megawatts.

A large number of manufacturers in key high energy consumption industries such as metallurgy, non-ferrous metals, chemicals and building materials, and even a host of medium-sized and large enterprises, still use outdated production processes, thus resulting in high energy consumption levels and seriously polluting the environment. Shanghai's Baoshan Iron and Steel Complex has an energy consumption level of 737 kg of standard coal for production of one ton of steel, meeting advanced world standards, but it is an exception. Another 26 large and medium-sized iron and steel complexes still have an energy consumption level of more than one ton of standard coal for production of one ton of steel.

Therefore, the use of advanced domestic and foreign technology to improve the technological standard of these enterprises is not only urgently needed for them to lower their energy consumption, improve their economic returns and enhance their market competitiveness, but also serves as an inevitable road for reducing environmental pollution at the root source.

In particular, the adoption of the existing mature technology for techno-logical revamping can lead to quick and high economic returns. For example, the metallurgical industry has raised its continuous casting ratio from 67.6 per cent in 1998 to 80 per cent in 2000; production of continuous rolled and semi-continuous rolled steel increased from 19.7 million tons to 26 million tons. In the chemical industry, 90 per cent of soda ash plants that had not previously pursued energy conservation revamping through ion membrane technology had begun upgrading themselves by using domestically built equipment, at the time of writing; and 200 small synthetic ammonia factories (with a total annual production capacity of four million tons) were adopting steam self-sufficiency technology.

The building material industry has developed new energy-efficient wall materials and compound materials that can utilize wastes, as well as non-metallic technological products involving intensive processing and with a high technological content and light pollution. By 2000 the sector had cut the existing production of 500 million solid clay bricks by 30 per cent, and increased the ratio of new wall materials to the total from 15 per cent to 35 per cent. The full undertaking of these projects alone requires a total investment of 30 to 33 billion yuan, but it can form an annual energy conservation capacity of 13 million tons of standard coal.

Because of an extensive production and operational mode formed many years ago, China has an enormous amount of abandoned resources in industrial production. Nevertheless, part of these resources can be recovered and utilised under the current technological and economic conditions. For instance, in the course of production metallurgical, chemical, petrochemical and building materials all produce large amounts of combustible gas and exhaust heat.

Large and medium-sized metallurgical enterprises have an average furnace gas emission rate of 11.3 per cent. The petrochemical industry emits nearly 900 million cubic metres of combustible gas each year. The power industry produces 120 million tons of coal ash a year. The coal industry turns out 130 million tons of gangue (rock or mineral matter of no value) and 60 million tons of coal mud annually.

These wasted resources have not only seriously polluted the environment, but also have wasted large amounts of energy. To tackle the problem, the metallurgical industry, for example, was in the process of bringing down the furnace gas emission rate to five per cent, and increasing the recovery of converter gas to 70 cubic metres for the production of one ton of steel. Meanwhile, 15 pressure differential power-generating units were being installed in major furnace ceilings at Wuhan Iron and Steel Company, Taiyuan Iron and Steel Company and the Capital Iron and Steel Company (Beijing). These have a total capacity of 99 mW and can generate 588 million kWh of electricity a year.

The chemical industry has agreed it could install exhaust heat and power generation facilities on 53 sulphuric acid production units with an annual capacity of 40,000 tons, to add a power generating capacity of 65.5 mW. It was also estimated that projects for comprehensive use of gas and for integrated production of heat and electricity from gangue could be undertaken at 60 medium-sized synthetic ammonia plants with an annual production capacity of more than 60,000 tons.

Moreover, through inter-industrial association, the use of coal ash and gangue to produce building materials, and coal mud and gangue for power generation also have high economic returns and good environmental effects. The undertaking of these projects was expected to require a total investment of 3.5 to four billion yuan, but would form an annual energy conservation capacity of 4.1 to 4.3 million tons of standard coal.

All these projects look good on paper, but will only succeed if the local political will is present and the companies involved are given enough incentives to make them work. The cost is a major concern, but could be offset through collaboration with foreign investors either providing cash or technology in return for an equity share in the business.

The threat of 'carmageddon'

As noted previously, fumes from vehicle exhausts are another source of danger not just within cities themselves. For the impact of growing car ownership, for example, also spreads beyond the urban areas, perhaps as 'carmageddon' as the geographer John Adams has put it. Vehicles require roads, steel, plastics, oil, aluminium, chrome and many other inputs, all of which put a strain on the ecosystem and the exhaust fumes vehicles produce have a regional not just an urban dimension. By 1998, China produced 1.6 million vehicles, including 0.5 million motor cars (China Statistical Information and Consultancy Centre, 1999, p. 12), thus adding to the 6 million trucks, 12 million motor cars and other vehicles already on the road in 1997 (State Statistical Bureau, 1999, p. 536) and contributing to the ecological tramplings motor vehicles cause. Table 8.6 shows how the volume of passenger traffic, for example, has grown since the early years of the PRC. As can be seen, the number of rail passengers was more than a billion by 1995, although this has declined slightly since then, while the number of road passengers was more than 10 billion by that date, and is expanding rapidly. Road passengers exceeded rail passengers in number by 1970, and in terms of

Table 8.6 Passenger traffic, 1952–99

Year	Rail passengers (million)	Road passengers	Civil aviation (million)	Rail passenger-kilometres (100 million passenger km)	Road passenger-kilometres (100 million passenger km)	Civil aviation passenger kilometres (passenger km)
1952	163.5	45.6	0.0	200.6	22.6	0.2
1962	750.0	307.4	0.2	859.0	141.5	1.2
1970	524.6	618.1	0.2	718.2	240.1	1.8
1980	922.0	2,228.0	3.4	1,383.2	729.5	39.6
1990	957.1	6,480.9	16.6	2,612.6	2,620.3	230.5
1995	1,027.5	10,408.1	51.2	3,545.7	4,603.1	681.3
1997	933.1	12,045.8	56.3	3,548.5	5,541.4	773.5
1999	1,001.6	12,690.0	60.9	4,135.9	6,199.2	857.3

Source: State Statistical Bureau (2001), *China Statistical Yearbook 2000*, Beijing: China Statistical Publishing House, Tables 15.6 and 15.7, p. 515.

passenger-kilometres the road total exceeded the rail total by 1990, also more than doubling by 1997. In the 1990s, civil aviation too has markedly expanded, with total passengers reaching the 50 million plus mark by 1995, and adding nearly the population of Scotland in the two years to 1997. Such growth, as in other countries, puts enormous pressure on the environment; in China of course it is the sheer scale of potential vehicle ownership that many find so perturbing.

Of course, it is not just passengers that are carried across China; freight is also of enormous importance. The details are shown in Table 8.7. The volume of passenger traffic carried by waterways is relatively low, with nearly 226 million passengers in 1997, and a total of almost 156 million passenger-km, and so is omitted in the previous table (State Statistical Bureau, 1999, p. 539). In Table 8.7, however, waterways come into their own, and here the data excludes that for civil aviation, which is (again, relatively) minimal (total tonnage of freight carried by air in 1997 was 1.2 million tons, and 29 million ton-km) (ibid., p. 540). As before, the increasing importance of road traffic can be seen, with total tonnage of freight carried by road outstripping that by rail by 1980 alone, at that time by a factor greater than three.

By 1997, as Table 8.7 shows, the factor was now six for total tonnage. At least, however, rail continues to hold its own for the longer haul, and remains more than twice the total ton-km of road, although the ratio has reduced markedly since the early 1950s, as the table exemplifies, when it was approximately 40:1. It does not take much imagination to picture, and indeed smell, the impact of this road traffic especially on the ecosystem, with diesel fumes (regarded as being highly carcinogenic) and other noxious substances being spewed out into the atmosphere. However, coal-burning trains, although more romantic than modern diesel or electric ones, are still

Table 8.7 Freight traffic, 1952–99

Year	Rail (million tons)	Road (million tons	Waterways (million tons)	Rail freight kilometres (100 million ton-km)	Road freight kilometres (100 million ton-km)	Waterways freight kilometres (100 million ton-km)
1952	132.2	131.6	51.4	601.6	14.5	145.8
1962	352.6	327.9	174.6	1,719.1	62.1	452.6
1970	681.3	567.8	254.4	3,496.0	138.1	931.3
1980	1,112.8	3,820.5	426.8	5,716.9	764.0	5,052.8
1990	1,506.8	7,240.4	800.9	10,622.4	3,358.1	11,591.9
1995	1,658.6	9,403.9	1,131.9	12,836.0	4,694.9	17,552.2
1997	1,720.2	9,765.4	1,134.1	13,046.4	5,271.5	19,235.0
1999	1,672.0	9,904.4	1,146.1	12,577.9	5,724.3	21,263.0

Source: State Statistical Bureau (2001), *China Statistical Yearbook 2000*, Beijing: China Statistical Publishing House, Table 15.8 and 15.9, p. 516.

found in China and these too, add pollutants to the atmosphere. Even waterway traffic, which has the highest figures for ton-km since at least 1990 (actually, more detailed examination of this Yearbook shows that 1988 was the first year when it outstripped rail), is by no means pollution free, and adds to the watercourse pollution discussed in the following section.

Waste disposal

Waste disposal is an issue that concerns every part of the mainland and beyond, as we have already discussed in Chapter 5 in regard to the disposal of household rubbish. In Hong Kong, waste is not just from the population itself but also from its wide range of industrial plants and port activities. From a visit to the EPD in 1995, for instance, the author Cook was told that population had grown 30% from 1980–95, GDP 300% but waste, too, had kept pace with this, also at 300%.Revealing British influence, the unit used was double-decker busloads! And so, each day, the equivalent of 480 double-decker busloads of garbage was generated, with 25 double-decker busloads of plastic bags alone. Hong Kong and Taiwan represent the affluent societies that the PRC is seeking to emulate, especially in the coastal region. However, as we have seen in this book economic development poses environmental dangers. The major study by the Chinese Academy of Sciences cited previously attempted to quantify this for waste discharge. Wang Yi, one of the two main authors, and his colleagues estimated that:

> [each] increase of one percentage point of the total industrial output value means a growth of 0.17 per cent and 0.44 per cent of industrial waste water and solid wastes. [Assuming annual economic growth of 6–7

Table 8.8 Forecast waste discharge, 1990–2000; actual discharge 1985, 1997, 1999

Year	Waste water (billion tons)	Waste gas (trillion cu. m)	Industrial solid wastes (million tons)
(1985	32.8	7.1	461.5)
1990	35.9	8.3	541.6
1995	38.5	9.8	614.2
[1997 actual	41.6	11.3	657.5]
[1999 actual	n.a.	11.5	649.1]
2000	41.3	11.5	693.5

Source: Data for 1985 and predictions for 1990, 1995 and 2000, Hu Anyang and Wang Yi (1992), *Survival and Development – A Study of China's Long-Term Development*, The National Conditions Investigation Group Under the Chinese Academy of Sciences, Beijing: Science Press, p. 153. Data for waste water 1997, Cook, I.G. and Murray, G. (2000), *China's Third Revolution: Tensions in the Transition to Post-Communism*, London: Curzon. Data for waste gas and industrial solid waste 1997 and 1999 from State Statistical Bureau (1999 and 2001), *China Statistical Yearbooks 1998 and 2000*, Beijing: China Statistical Publishing House, p. 802 and p. 770 respectively.

per cent] the discharge of [total] waste water will amount to 41.29 billion tons by the end of the century; waste gas, 11,500 cubic metres; industrial solid wastes, 690 million tons, representing 18.3 per cent, 49.4 per cent and 31 per cent increase over 1987 respectively.

(Hu Anyang and Wang Yi, 1992, p. 153)

Table 8.8 details these estimates for 1990–2000 (1985 data was already available at the time of the study). What actually happened was that average growth rates in the 1990s have generally been higher than predicted and so by 1997 alone, total waste water was already over the 2000 prediction, at 41.6 billion tons (Cook and Murray, 2000, op. cit.), waste gas was also greater than predicted, at 11.3 trillion cu. m by 1997 (cf. the estimate of 11.5 trillion cu. m for 2000) and the quantity of industrial solid wastes was 657.5 million tons in 1997, well on the way to the 2000 forecast (State Statistical Bureau, 1999, p. 802). The economic downturn of 1998, however, had the paradoxical effect, as in other countries, of slightly improving environmental quality via reduction of, for example, waste water to 39.5 billion tons in all (Li, 1999, p. 16), although SEPA cite a figure of 800 million tons of industrial solid waste for that year, which does seem out of line with that given by the State Statistical Bureau for 1997 (ibid.).

In our previous book we cited concerns about the ecology of the Yangtze, for example, with Chongqing discharging 70% of the pollution of the entire basin, for much of its sewage is pumped untreated into the river and in addition, only 40% of the city's industrial waste reaches national standards for disposal (Cook and Murray, 2001, op. cit.). The city had no waste disposal plant whatsoever until after it was taken under direct control of the central authorities in 1997. This is an extreme example, but China generally cannot keep pace with the volume of sewage and industrial waste discharged into

Box 8.3 Wastewater pollution – three case studies

As was noted in the last chapter, Tianjin is one the mainland cities suffering a serious water shortage. One effect of this has been to force farmers to use untreated wastewater (municipal combined with industrial) for agricultural and drinking purposes. But these municipal and industrial waters contain chemicals, organic, and thermal wastes, and sewage consisting of human waste that can pose a serious health threat. Tianjin is the third largest industrial city in China. For many years, the construction of urban drainage, sewage, and wastewater treatment (domestic and industrial) has lagged far behind the development and environmental requirements for the city. A Wastewater Improvement Project has been introduced and co-financed by the Tianjin Municipal Government and the World Bank (providing US$400 million) to help improve conditions. This project focuses on sewage collection and treatment, as well as measures to improve the institutional controls on sewage and pollution control. To ensure that the Tianjin Municipal Government achieves its overall objectives of optimising water consumption and reducing pollution loads, an emphasis on the reduction of industrial water consumption and the improvement of effluent discharge quality. The entire area will cover about 240 sq. km. 1788 kms of interceptors and main sewers within the city boundary, 131 pumping stations, 171 kms of river embankments, and two sewage treatment plants (author research).

Meanwhile, 300 miles west of Shanghai, lies the Chao Lake, one of China's five largest freshwater lakes. Located in Anhui Province, the lake serves Hefei, the provincial capital (population 1.1 million), and Chaohu City, (population 170,000) for irrigation purposes, and is also the main source of potable water. In recent years, the rapid development of industrial and urban areas, and changes in agricultural practices, have caused damage to the lake. Although pre-treatment or final treatment of industrial waste water has controlled pollution by heavy metals, toxic and hazardous substances, the lake has become overloaded with nutrients from municipal and industrial wastewater due to soil erosion and from excessive applications of chemical fertilizers on agricultural land. As a solution to restore the lake's water quality and to improve the health and economic concerns for the area, The Asian Development Bank (ADB) appointed the consultants Camp, Dresser & McKee (CDM) to conduct a feasibility study to develop a possible project for the construction of municipal waste water treatment facilities (CDM, 1999).

Meanwhile, Xi'an, capital of Shaanxi Province, with a population of 2.8 million within the urban area, has had to deal with major water pollution problems. There are many industries in the area, and vast quantities of untreated domestic and industrial wastewater have been flowing into small streams leading to the Wei River, and this has seeped into the city's groundwater system, which is used for irrigation, industry, drinking, and rural home and stock watering, raising environmental and health concerns for the population. To cope with this, work began in 1997 on developing the Beishin Qiao Wastewater Treatment Plant, due for completion by 2001. (Krüger, 1997)

water courses, hence the 100 million tons figure for untreated sewage noted by Huang above (see Box 8.3 for further discussion of the wastewater problem at three specific mainland locations and Box 8.4 for a delightful glimpse of one aspect of the sewage disposal problem related to poorer districts of Shanghai).

Also, by 1997, for instance, only 62% of industrial waste water nationally reached the standard for discharge (State Statistical Bureau, 1999, p. 802). There is discrepancy between SEPA and the State Statistical Bureau data for treatment of industrial waste water, but even the higher figure of nearly 85% of industrial waste water treated in 1997 (given by the State Statistical Bureau 1999, p. 802, cf. SEPA's 79%) would still leave approximately 3.8 billion tons of industrial waste water untreated.

As regards water consumption, half of China's cities face water shortages, to such an extent that urbanites were advised in 1998 not to leave the tap running, and to 'Listen during the quiet hours for water running beneath the flush toilet' (Zhao Huanxin, 1998)! This was to save water via cutting waste and fixing leaks, to such an extent that water saved by 2000 would be able to meet half of the added requirement for urban industrial water by that date. In Beijing's hotels, for instance, there are notices over the taps asking visitors not to use water unnecessarily, while the more sophisticated hotel urinals have infra-red controls on the flushing mechanism, sensing when someone is about to use them, rather than flushing at arbitrary intervals.

In all, waste disposal is a difficult, messy and sometimes dangerous business. SEPA suggests, for example, that there were 10 million tons of dangerous industrial solid waste to dispose of in 1998 alone (Li, 1999, op. cit.). With this topic, as for others covered in this book, the PRC has a long and difficult road to travel in order to develop and conduct effective policies. In the next chapter we shall consider some of the recent successes and failures of environmental policy across all regions ofChina.

Box 8.4 Farewell to potties

With a chorus of rhythmic clattering, women line up in front of the public toilets to dump the contents of their matongs, the wooden chamber pots used by their families in the absence of modern plumbing. This is a scene that is repeated every day in many of Shanghai's old residential areas. For decades, millions of Shanghainese have been using chamber pots at home because of poor housing and sanitation conditions. In fact, chamber pots are a household necessity in many southern provinces.

Unlike northerners, southerners think it too much of a hassle to get dressed at night and trek all the way to the public toilet. Living without a flush toilet, they find the comparative convenience of the chamber pot a better choice. Many Shanghainese dream of getting rid of their matong and moving to a spacious apartment with modern plumbing. Since the late 1980s, dilapidated houses have been swept away on a wave of residential high-rise construction projects. Every year, tens of thousands of people bid farewell to their chamber pots and move into new homes with en-suite bathrooms.

Grandma Kang Guihua, 70, has been emptying her chamber pot every morning for more than half a century. Not far from her home in the northeast part of the city, there is now a new skyline in the form of a residential village called Lucky Garden. Many of its residents have been relocated from an area near the Bund to make way for a new business and financial plaza. 'We are not so lucky,' said Grandma Kang, who lives with her son, her daughter-in-law and her two-year-old grandson in her 30 sq. m home. 'It is unlikely that this place will become a new development site. Anyway, we are frantically saving money to buy our own new home.' An apartment at the cheap end of the market is at least 2500 yuan per sq. m. A simple two-bedroom apartment costs at least 150,000 yuan, while an average worker earns less than 1000 yuan a month in Shanghai. 'We cannot afford it right now', she lamented. 'So, we have to keep using the matong.'

But even the matong is being modernized. In the same neighbourhood, young wives use a smaller enamel chamber pot instead of the old wooden one. 'It is more than 30 years old, older than you', Grandma Kang said while pointing to her empty matong outside in the sun. 'It is a thing of the past. I hope I can live to see the day when I can throw it out with the trash'. (Shi Hua, 1999)

9 Pollution on the periphery

It is not just the traditional mainland area of the PRC that faces environmental problems; Hong Kong and Taiwan, too, have a range of environmental issues with which to contend. Meanwhile, the PRC's most controversial piece of claimed territory, Tibet, has unique environmental problems as its own – some created by its unique geography, others by the political system introduced since the PRC takeover in 1950.

Tibet has always been a land of extremes – of cloud-capped mountain peaks and lush tropical valleys – held together in a delicate mix of ecosystems that inspired a mystical reverence for nature and life. To generations of Buddhist monks, nomads, and peasants alike, their homeland was a dream-like natural paradise. But the march of Chinese troops into the remote Himalayan region, and the subsequent tapping of Tibet's resources to feed China's economic development, has made some Tibetans fear their fragile world could be forever thrown out of balance. Myriad serpentine roads, for example, have transformed the once virtually impenetrable region, allowing an inflow of Han Chinese migrants, and the outflow of trucks carrying off lumber and other resources.

It is very difficult to obtain a truly accurate picture of what is happening in Tibet regarding both damage to and protection of the environment. The Tibetan government in exile, set up when the Dalai Lama fled the country after anti-Chinese riots in 1959, tends to present a uniformly gloomy picture of rape and pillage, which is generally backed by Western environmental and human rights groups. The Chinese Communist Party, meanwhile, tends to wax lyrical about the wonders wrought by brave Han Chinese pioneers in wrenching Tibet from the mire of medieval feudalism to enjoy the full benefits of socialism in the great Chinese family. In what follows, we briefly set out the two cases and attempt to draw some reasonable conclusions.

Tibet: Land of Riches

Tibet's complex topography and widely varying climates has resulted in an abundance of natural resources. Rivers offering enormous potential water-power criss-cross its 1,220,000 sq. km area. The continuous snow-covered

Table 9.1 Tibet's plant and animal diversity

	Number of species/coverage	Key Examples	Location
Plants	5,000+	1,000 medicinal plants; 200 edible fungi	Gyirong, Yadong and Zhentang in west, Medog, Ziyu and Lhoyu in southeast
Forest	9.84%	Himalyan fir, pine and spruce, alpin and Tibetan larch, cypress, Chinese juniper	Humid sub-alpine zones of Himalayas, Nyainqengtanglha and Hengduan ranges
Animals	142 mammals, 473 birds, 49 reptiles, 44 amphibians, 64 fish, and more than 2,300 insects	Assamese macaque, rehsus monkey, muntjak, deer, antelope, serows, snow leopards, wild cats, little pandas, wild donkeys, lynxes and blue sheep	Widely spread

Source: Various goverment reports of Tibetan Autonomous Region.

mountains and valleys and the North Tibet Plateau house a wide variety of minerals. The eastern and southern parts of the region are largely covered with primeval forests, home to rare animals and plants (see Table 9.1).

It is often subdivided into 'inner' or 'eastern' Tibet, and 'outer' or 'western' Tibet, or more environmentally into the inhospitable Northern Plains – 'the land of no man and no dog' – a bleak, monotonous grassland region of nomadic herders dotted with many saltwater and freshwater lakes, but containing valuable mineral and geothermal resources, Southern Tibet with its high range in both altitude, diurnal and annual temperatures, and high winds, where animal husbandry is practised on the upland meadows and subsistence agriculture in the valleys, and Eastern Tibet which is the most fertile part, with forests, grassland and cultivation of such crops as highland barley, winter wheat, rye and garden vegetables (Tiley Chodag, 1988).

There are more than 90 known mineral types found in Tibet, and Chinese mining geologists have already determined significant reserves of 26 of these, with 11 ranking among the top five by province. Chromite deposits, concentrated along the Lake Pangkok Co to the Nu River rift zone in northern Tibet, and along the Yarlung Zangbo River rift zone, are the largest in China. Prospective lithium deposits are among the largest in the world and Tibet is already China's key lithium production base. Prospective copper and gypsum reserves rank second in the country; boron, magnesite, barite and arsenic third, mica and peat fourth, and kaolin fifth. Other significant mineral deposits include salt, natural soda, mirabilite, sulphur, phosphorus, potassium, diatomaceous earth, corundum, rock quartz and agate.

Tibet has been regarded as weak in traditional energy resources such as coal, oil and natural gas, but rich in hydro, geothermal, solar and wind energy. It produces approximately 200 million kW of natural hydro energy annually, about 30 per cent of the national total. The main stream of the Yarlung Zangbo River has an estimated exploitable natural hydro energy capacity of 80 million kW a year. Chinese engineers have found more than ten sites and sections of the river suitable for the construction of hydropower stations. At one point, the river curves through a spectacular 200-km long gorge while dropping 2,190 metres. Experts believe that a 36 km channel could be cut through the rock to channel the water to a point where a giant 40 million kW hydropower plant could be built.

Meanwhile, more than 600 geothermal sites have been located in the Nu-Jinsha–Lancang tectonic zone, the Yarlung Zangbo rift zone and the Nagqu–Nyemo rift zone, including hot springs, boiling springs, geysers, hot-flow rivers and exothermic ground surfaces, with an estimated heat discharge of 550,000 kilocalories per second, the equivalent in heat produced annually to about 2.4 million tons of standard coal.

However, it does seem that oil and natural gas are present under the Tibetan plateau and could be exploited. In 1999, the China National Star Petroleum Corp (CNSPC) reported it had found the region's first oilfield with initially proven oil reserves of three million metric tons. The company said that the deposit, in the Lunpola Basin, 300 km north-west of Nagqu in northern Tibet, might contain up to 10 million metric tons, and that the total oil and natural gas resources in the basin as a whole could range from 100 million and 150 million metric tons

Exploiting such natural resources, the Chinese Government would surely argue, would bring immense economic benefits to the Tibetan people. However, the Tibet government in exile would undoubtedly counter with claims of the looting of the region's treasures for the sole benefit of the Han Chinese majority (see below), without even beginning to go into concerns about the environmental damage that would be caused by excessive mining and tampering with the region's fragile ecology.

Tibet: the Chinese case

The central government in Beijing is proud of its achievements in Tibet since the early 1950s, through special economic support to raise living standards and haul the region out of 'feudal slavery' (see Chapter 6 of Cook and Murray, 2001). It estimates that, over the last 40 years, it has spent more than 35 billion yuan in Tibet on financial subsidies and investment in key construction projects.

By Chinese standards, and probably those of most of the world, Tibet was backward. Few of its resources were being tapped – partly due the tenets of the Buddhism the people devoutly observed – and the majority of the population lived in conditions close to poverty. In the literature of Chinese

Communism related to Tibet, therefore, its 'peaceful liberation' in 1950 and the commencement of 'democratic reform' to sweep away the last vestiges of 'feudalism' when the Dalai Lama fled the country in 1959 are joyful events widely welcomed by the liberated Tibetan populace. The development of commerce and industry, and the construction of a modern infrastructure, therefore, can only be regarded as positive steps to haul the region out of medieval backwardness and into the twenty-first century.

In the first half of 1994, the Chinese government decided to invest an additional 2.38 billion yuan in 62 engineering projects regarded as urgently needed for Tibet's economic development. Their great importance, government officials repeatedly stressed, lay in improving the region's backward infrastructure, raising living standards and providing a solid foundation for development in the new century.

Thirteen of the 62 projects involved agriculture and water conservancy (24.8 per cent of total investment), 15 energy (27.3 per cent), seven transportation and telecommunications (9.2 per cent), six industry (7.3 per cent), and 21 social services and municipal construction (31.4 per cent). The state provided three-quarters of the necessary investment, with the remainder coming from 29 non-Tibetan provincial-level governments and six major cities, each contributing aid to particular projects according to their ability to do so.

Modern industry first began in Tibet in the late 1950s, with the appearance of more than ten new fields of production, including electric power, mining, wool spinning, forestry, food processing, printing, building materials and machine processing. Today there are more than 260 medium- and small-scale state-owned enterprises, employing 51,000 workers.

Massive efforts have also gone into overcoming Tibet's remoteness, through the construction of all-weather highways from various western mainland provinces such as Qinghai, Sichuan and Yunnan as well as with neighbouring Nepal. The final missing link will be removed with the construction of a railway from Golmud in Qinghai Province to Lhasa (of which more later).

Throughout the 1990s, a great deal was done to develop a more 'industrialised' modern agricultural sector, and this effort is continuing into the new century. A government report described how in the agricultural areas of central Tibet, grain production was being stressed, while forestry and cultivation of fruit and vegetables were also being developed: 'Integration of agriculture and animal husbandry are being pursued with the goal of producing more than enough grain, meat and milk for self-sufficiency. Irrigation-based cultivation is being strengthened, as is basic construction in the pasturelands. The acreage of land with reliable irrigation will be expanded'.

Changing agricultural patterns

The 'Three Rivers' Comprehensive Agricultural Development Project is the big one, concentrating on areas around the middle reaches of the Yarlung

Zangbo River and its two tributaries, the Lhasa and Nyangqu. It is the largest project both in terms of scale and investment, affecting 65,700 sq. km housing 36 per cent of the Tibetan population. In the ten years from 1991, a billion yuan was invested in building the area into a production base for commodity grain, animal by-products, light industry and food processing. The eventual targets were an increase in annual production of 150 million kg of grain and 24 million kg of meat.

The project also involves heavy investment in creation of modern power generation facilities. One Chinese reporter noted that what was most interesting on his visit to a remote pastureland was to find that most herders had abandoned their former nomadic life to move into new houses. He described walking into one house and discovering electric light bulbs, only to be told by a county official accompanying him that 'We don't have a power station in the county. Herders have them for decoration only' (Yan Zhenghong, 1999).

Before 1950, the Togde Power Plant built by the Tibetan Gaxag government in 1928 was the only one of its kind in the region. Located in the Togde Gully in Lhasa, it had a meagre installed capacity of 92 kW and served mainly the Potala Palace, the Lhasa Mint and the small number of noble families. It was damaged by flood in 1944. By 1990, 15 county-level hydro-electric power stations had been built in the Three River Valley area to generate 3.9 billion kWh of electricity a year, with the formation of three power grids centred on Lhasa, Zetang (in Shannan) and Xigaze. This, the government propagandists say, will not only enable the programmes for agricultural modernisation to be achieved, but will also transform the lives of Tibetan peasants.

This is the key issue. The rangelands of the Tibetan Plateau and adjoining Himalayan region are one of the world's great grazing land ecosystems. Stretching for almost 3,000 km from west to east and 1,500 km from south to north, the region is one of the largest and most important pastoral areas on earth. The fact that these grazing lands have supported pastoral cultures for thousands of years while sustaining a varied and unique flora and fauna bears witness to the existence of a remarkably diverse and resilient rangeland ecosystem.

These grazing lands also form the headwaters environment where many major rivers have their beginnings – the Yellow, Yangtze, Mekong, Salween, Brahmaputra, Ganges, Indus, and Sutlej rivers. The preservation and management of these river source environments have global implications, as the water from their watersheds will be of increasing importance in the future. Upsetting the ecological balance in these high-elevation rangelands will have a profound effect on millions of people living downstream.

In recent decades, however, many changes have taken place that are transforming traditional rangeland use and conditions, pastoral systems, and the lives of herders dependent on rangeland resources. Nomads and their pastoral systems have always been confronted with events that change their lives – droughts that wither grass, winter storms and livestock epidemics that wipe out herds, and tribal wars that displace people and their animals – but the changes they are facing now are likely to have more significant, long-term implications on their way of life and the ecosystems in which they reside than any previous changes.

Such new changes include the modernisation process itself, which as already outlined has brought improved access and services to previously remote nomadic areas and increased demand for livestock products; the expansion of agriculture onto rangelands and decrease in the amount of grazing land available to nomads' herds; disruption in traditional trade networks, which were often an important part of pastoral systems; the expansion of the protected area system, with increased regulation limiting livestock grazing; and, more recently, policies to settle nomads and divide rangelands into individual family parcels.

In many cases, the changes have altered previous, often stable, relationships between pastoralists and their environment. Pastoral systems are still in a state of transition and it is not yet clear what patterns will eventually emerge.

With the increase in human population in the region, along with a rise in peoples' incomes, there is an increasing demand for livestock products from pastoral areas. Many nomads have now entered the market economy, selling their livestock products and purchasing goods they require, in contrast to traditional barter systems. Many pastoral families have greatly improved their standard of living. Nomads throughout the Tibetan pastoral areas of western China, who until a few years ago still lived in tents the year-round, have now built houses and barns, and have erected fences around private winter pastures, although most herders continue to live in tents in the summer.

Herders are also demanding improved social services (schools, health clinics, etc.), as well as improved veterinary services and market outlets for livestock products. Keeping abreast of the changes taking place on the grasslands is an important task for pastoral researchers. These changes and the effects they have had – and are having – on the rangelands, livestock production and socio-economic dynamics of pastoral societies, need to be closely monitored in the years to come before deciding whether they are, as the Chinese claim, for the long-term good of Tibet and its people.

Tibet: the exiles' case

Certainly, these environmental aspects, as well as most of the economic-related policies and practices introduced by the Chinese since 1950, are not welcomed by the bulk of the Tibetan exiles scattered around the world. A study submitted by the Tibetan government in exile to the Eighth Session of the UN Commission on Sustainable Development, for example, claimed that Chinese policies had led to widespread environmental damage on the Tibetan Plateau and were 'of no benefit to the Tibetan people' (Tibet, 2000). China was only interested in grabbing Tibet's natural resources for its own advantage and, in the process, is destroying an ancient lifestyle and culture through environmental degradation and population transfer of Chinese settlers, it claimed. The study's key findings can be summarised as follows:

Damage by mining. Ecologically catastrophic mining, which had earned China over US$2 billion from 1952 to 1990, would continue to escalate. Mineral reserves found in U-Tsang (Central Tibet) alone were valued at US$81.3 billion. The Tsaidam Basin, which covers an area almost the size of Britain, had oil reserves of 42 billion tons and natural gas reserves of 1500 billion cubic metres. All of the gas reserves would be utilized to supply China's booming industrial cities, like Shanghai, Wuhan and Nanjing.

Damage by deforestation. As already discussed, Tibet is the planet's largest and highest plateau and has ten major rivers flowing from its glaciers. These rivers sustain 85 per cent of Asia's population, which is 47 per cent of the world's population. Widespread logging has led to heavy siltation of these rivers. The felling of 46 per cent of pre-1950 forest cover has led to growing desertification and floods in China and south Asia. Reports from the World Watch Institute estimate that the heavily forested region from Amdo (Qinghai) to the Yangtze River Basin has lost 85 per cent of its original forest cover. As a result, the Yangtze now discharges 500 million tons of silt a year into the East China Sea, a volume equal to the total discharge of the Nile, Amazon and Mississippi rivers combined. In 1998, the Brahmaputra, too, saw unprecedented floods in the Indian subcontinent (as did the Yangtze, as we have already discussed). Landslides and soil erosion caused by deforestation have increased the silt flow into the Bay of Bengal. One third of the two billion tons of sediment is deposited in the plains of Bangladesh, reducing the depth of rivers and causing disastrous floods every year.

Damage caused by hydroelectric projects. Major dams and reservoir projects under construction may solve the power shortage crisis in China, but will destroy the ecological stability of Tibet. China is permitting a private corporation to build, own and operate hydropower dams, which will only displace Tibetan nomads and not generate any electricity for Tibetans.

Damage from pollution. The once pristine waterways of Tibet are now polluted by chemical, nuclear, agricultural and industrial waste. In 1971, the first known nuclear weapon was brought into Tibet. Today, China has 17 secret radar stations, 14 military air bases, 8 missile bases, 8 ICBMs, 70 medium-range missiles and 20 intermediate-range missiles in Tibet. To support its nuclear programme, China has established uranium-mining sites in many regions of Tibet, adding an even more dangerous component to the existing water pollution problems of Tibet's waterways. Tailings from large-scale mining operations are a primary source of water pollution today in Amdo. Rivers around Lhasa already report mounting pollution problems from untreated sewage, industrial waste, and salts and nitrates leaked from fertilizers used in intensive farming projects designed to meet the food needs of Central Tibet's expanding Chinese population.

Damage to biodiversity. A biodiversity comparable to the Amazon Rain-forest in its richness is endangered by China's economic exploitation of Tibet, even before all of it has been documented. The massive deforestation and mining has accelerated the destruction of the fragile, once untouched, environment. Rare species of animals, birds and plants are now on the brink of extinction. There are at present 81 known endangered animals on the Tibetan plateau. Its pristine ecology is being destroyed by a pumped-storage plant to supply Lhasa's electricity needs – a project whose design is now judged to be faulty and leading to lowering water levels, increased salinity and habitat loss for diverse and rich wildlife, including birds and fish. Mammals are largely endangered due to hunting and poaching, some for the commercial value of their wool, antlers, skins, fur, bones, and inner organs; some gunned down as trophies to take home to China or sell as meat. Fish are dynamited in lakes and rivers. While wildlife conservation laws are in place in the legal system, their implementation is weak and barely enforced since wildlife is a state-controlled commodity and therefore categorized as a renewable and exploitable resource.

Damage to health. It is known that China still employs shallow burial techniques for nuclear waste, a method now obsolete in the West, and remote regions of Tibet are earmarked in Beijing's plans to trade in the profitable recycling of hazardous and toxic wastes from developed nations. Already an abnormal rate of childbirth mortality, birth deformities, unprecedented and mysterious illnesses in humans, and high death rates among animals and fish, are recorded from regions around two nuclear production departments in Amdo. Nomads and villagers around the Ninth Academy also experienced high rates of cancer in children, similar to findings in Japan post-Hiroshima.

Apart from the claim about environmental destruction from the 'use of Tibet for the production of nuclear weapons and dumping of nuclear waste' (Jetsun Pema, p. 210), Tibetan exiles also claim 'Massive deforestation in eastern Tibet, big game poaching, over-exploitation of mineral and other natural resources' (ibid., p. 232). A two-day public hearing on Tibet's environment, held in Dharamsala on 16 and 17 November 1998, reflected such concerns among the officials of the exile Tibetan administration and other individuals.

For example, Thonsur Lobsang Tenzin from the Private Office of His Holiness the Dalai Lama opened the discussion by speaking on the state of Tibet's environment prior to the country's invasion by China's People's Liberation Army. Drawing on the influence Tibetan Buddhism wielded in promoting respect for the environment among its people, he spoke on the harmony that existed between the environment and the inhabitants of the Tibetan plateau (Tibetan Bulletin, 1999).

Since then, as Tegyam from the government's Research and Analysis Department claimed, Tibet has been a miserable site for one of the world's most alarming rates of environmental destruction. 'If you look up at the sky in Tibet these days, for hours you will not be able to spot any birds', he said as illustration. He also stated that the 'drong', a wild yak, which used to be

Box 9.1 Damage by over-population

Incorporating Tibet into China's economic development programme has intensified the migration of Chinese to the plateau, further marginalizing and impoverishing Tibetans. Tibet's population has more than doubled as a result of both military and civilian Chinese immigration.

Jetsun Pema (1997, p. 132), a sister of the Dalai Lama who was later to become Minister for Education in the Tibetan government in exile, led the third delegation of exiles in an official visit to Tibet in 1980. In her autobiography she cites estimates of 7.5 million non-Tibetan settlers plus 300,000 Chinese troops across the whole of Tibet (including the area which is now part of Qinghai province). These figures contrast markedly with China's 1990 census which gave a total of just over 1.5 million Han in all Tibetan-inhabited areas, including those in Sichuan, Qinghai, Gansu and Yunnan, and only 80,800 in Tibet proper (Zhang Tianlu, 1997, p. 14).

Of the latter, Ovchinnikov (1995, p. 85) claims that 'Nearly half of them are residents of Lhasa. They are generally builders, doctors, and teachers working on a contract basis.' By year-end 1997, official data listed the population of the latter autonomous area as being 2,393,000, of which 2,324,000 belonged to the minority nationality, i.e. Tibetan or other minorities, leaving only approximately 69,000 from the Han (State Statistical Bureau, 1999, p. 42). The central government in Beijing repeatedly insists that most of the Han in Tibet are actually there on short-term contracts (as experts or with PLA units) rather than long-term residents.

As with other Tibetan issues it is nigh impossible to reconcile such contrasting data. We do know that it would be very difficult for high numbers of Han to live effectively in the rarefied altitude of Tibet, due to medical side-effects such as heart disease, multiplication of red blood cells, and pulmonary oedema, but the official Chinese data seems as much understated as the official Tibetan figures seem overstated.

plentiful, is now nearing total extinction. Similarly, an Indian official pointed out that the recent (1998) floods along the Yangtze (see Chapter 6) 'has its cause in the extensive lumbering in Tibet' (ibid.).

At a New Delhi Seminar in April 2000, Sinologists and environmentalists also expressed deep concern over what they see as 'militarisation' of Tibet, saying it posed a grave threat to its environment and could become an irritant in India-China relations. Tsultrim Palden, a campaigner for the restoration of Tibetan ecology, said that, as the forest cover decreases, the heating mechanism of the plateau diminishes, resulting in alteration of pressure systems, which either delay or reduce the Indian monsoon. This lingering snow cover disrupting the Indian monsoon has the potential to foster disasters for Indian agriculture, he warned. He described Tibet as, 'a dumping-pit for China's nuclear waste, which can result in a catastrophe affecting several generations' (PTI, 2000).

As Chapter 6 shows, deforestation upstream has been identified as being part of the cause of recent severe floods, although the official line has been to focus on Sichuan and Yunnan tree-felling, rather than Tibet. But, given that 'Tibet has one of the largest forest areas in the country [China], covering over 94.8 million mu' (Zhang Tianlu, p. 136), plus Western newspaper reports of large-scale timber felling dating back some years, it is highly likely that deforestation has indeed reached high levels in Eastern Tibet especially. Even Chinese sources themselves are beginning to admit serious environmental problems in Tibet, although the blame to us may seem misplaced. For instance, Zhang Tianlu cites 'The shortage of electricity and the resultant insufficient irrigation . . . for degradation and desertification of grasslands, reduction of soil fertility and deterioration of the eco-environment' (ibid., p. 138). With electricity, all will be well!

Similarly, the threat to the rare Tibetan chamois in the vast Hoh Xil reserve in west Qinghai is firmly blamed on the illegal hunting, the 'cruel' hunters themselves, and the high prices which their fine wool fetches as ring shawls in London and Paris, said to be anywhere from US$5,000 to $30,000 (Huang Wei, 1999, p. 17). These are serious reasons, but the lack of serious investment in anti-poaching activities or the lack of alternative sources of income in these interior provinces is not considered, even though the accompanying interview with Tsega, Director of the Reserve's Administrative Bureau, quotes him as noting that each of his young staff 'is in charge of protecting animals in an area of thousands of square km. Short of funds and equipment, they face great difficulties in accomplishing their task well' (*Beijing Review*, 5 July 1999, p. 20).

Another concern that will undoubtedly draw vociferous protests in the years ahead is the Chinese plan to improve Tibet's transportation links by building the region's first railway line. The project was first mooted in the early 1950s, following the immense logistical difficulties encountered by the PLA forces in 'liberating' Tibet, but repeatedly deferred due to the technical difficulties involved in lifting a railway to an elevation of 4,000–5,000 metres and through areas of permafrost.

The idea was revived in the mid-1990s and four routes were eventually identified. In late 2000, the government finally opted for one that would link the existing railhead in Golmud, Qinghai Province, with Lhasa, covering a distance of 1118 km. About 564 km of the railway will run inside Qinghai Province, and 516 km in the Tibet Autonomous Region. About 930 km will be at an elevation of over 4,000 metres, with the highest point at 5072 metres. Bridges and tunnels will comprise 30.6 km, and 552 km of the railway will cross an area of permafrost. Construction will take seven to eight years, with an estimated total investment of 19.4 billion yuan.

The project raised immediate concerns in the West, with some pro-Tibetan groups seeing it as a worrying extension of Chinese control over the disputed area, while environmentalists fretted about the possible ecological damage. The government in Beijing dismisses all such protests. But one major concern for Chinese railway planners certainly was how to deal with possible damage to the permafrost.

Throughout January and February 2001, a team from the No. 1 Prospecting and Designing Institute of the Ministry of Railways and the Chinese Academy of Sciences undertook a detailed survey of a key area. An experimental ground was set up at an elevation of 4,750 metres to test the rail bed, bridges, tunnels, housing, water supply and drainage facilities. Through repeated tests, these experts claimed to have found a way to solve the permafrost problems by controlling the height of the rail bed, building up an insulation layer, and erecting overpasses in certain areas. They concluded that the permafrost problem could be solved so long as the height of the rail bed reached 100 cm to 200 cm. Only time will tell if they are right, or whether the seeds of another ecological disaster have been sown for Tibet.

But, what of the overall future for Tibetan ecology?

> Although Tibet's natural environment is very harsh, the region is splendid and beautiful in its landscape and endowed with rich natural resources. Moreover, Tibetan people are hard-working and intelligent. So long as a workable strategy can be formulated based on the actual conditions for the social and economic development and implemented, a prosperous and affluent Tibet can be expected.
>
> (Zhang Tianlu, 1997, op. cit., p. 140)

It will, unfortunately, be many years before such a 'workable strategy' can evolve. Until then, the degradation of Tibet's environment will undoubtedly continue apace.

Prospects for reconciliation

Given the difference in cultural values between the Tibetans and the Chinese with regard to natural resource use, the most effective way to protect the region's environment arguably would be to restore Tibetan control of Tibet. This is very unlikely to happen, so the way forward has to be to find ways to influence China's management policies for the long-term good. This in itself is a complex topic, however, due to China's insistence on seeing any initiative from outside as a (Western) attempt to tear Tibet away from the 'loving embrace of the motherland'.

One possible way forward, satisfying both Chinese and Tibetans, however, is suggested by the success of the Qomolongma (the Tibetan name for Mt Everest) Nature Preserve, which is remarkable not only because it is restoring the environment by returning control over the land to local Tibetans, but also due to its curious coalition of backers. The project was conceived by the Chinese government and the US-based environmental group Future Generations, after seeking ideas and feedback from the area's authorities and impoverished peasants.

A decade ago, the region surrounding the planet's highest peak was a study in contrasts, from the pristine state of its summit descending to a garbage-strewn base. 'Nearby hillsides were being stripped of tree cover, and wanton hunting was jeopardizing the wildlife', says Daniel Taylor-Ide, President of Future Generations. The Tibetan, Chinese and American partners who a decade ago designed the preserve, roughly the size of Denmark, employed a concept at first alien to Beijing's central planners: a 'community-based approach to sustainable development', says Taylor-Ide (Platt, 1999).

He and other backers of the program began travelling from village to village, county to county, to ask farmers and hunters to join a massive joint venture. In exchange for limiting the destruction of forests and wildlife, the participants would receive, over time, access to primary health care, education, and energy-saving technology. 'Because the financing for the preserve came from the central government, Future Generations, and local villages themselves, each farmer realized he had a stake in the success of the program', says Taylor-Ide. The preserve resembles three concentric rings: the core is strictly protected from development, a buffer zone allows some use of natural resources, and an outer zone is designed to accommodate towns and business enterprises. Villagers and local authorities together volunteer to police protected forests and habitats, and divert funds that would have paid for preserve wardens into basic social services.

That approach toward sustainable development is empowering not only entire communities, but also ethnic minorities, women, the poor, and other groups often discriminated against. The remarkably simple idea has since taken off, and a few model villages have been copied throughout southern Tibet. Although Tibet remains one of the poorest regions in the world, Future Generations says life for the 75,000 residents of the Mt Everest preserve is showing steady signs of improvement. For example, glass windows and other low-cost, energy-saving materials are cutting fuel consumption, and deforestation has been sharply reduced. The area's wildlife is making a comeback and sales of many endangered species have been banned. 'The United Nations Development Program has launched a micro-credit program to finance individual businesses, and other aid groups are providing matching funds to build schools and roads' (UNDP, 1999).

Yet the community-based model, with its implicit embrace of people power and egalitarianism in decision-making on issues of development and conservation, still faces much resistance within the Chinese bureaucracy, say many aid workers. 'The Chinese are control-oriented folks', says Michael Rechlin, an expert on sustainable development of forestry and other natural resources at Paul Smith's College in New York. 'So it's not going to be an easy sell to extend community-based development to other areas', he adds (Platt, op. cit.).

With the Qomolongma Nature Preserve, he says, 'Future Generations and other international groups are building Tibetans' capacity to manage their own affairs'. In giving the area a degree of autonomy in mapping out a balance between economic and ecological progress, 'China wants to promote an environment that is not conducive to rebellion or discontent', Professor Rechlin says (ibid.). If that approach is adopted throughout the region, it could ease the five-decade-long clash of civilizations.

Hong Kong struggles to clean up

Hong Kong's Environmental Protection Department (EPD) built up an enviable reputation for its educational policies and legislation, but even before the handover struggled to keep pace with the drive for economic development in the colony, with the 'typical problem' of 'conflict between development and environment' (Hung, 1994, p. 254). The controversy over the new airport on the previously quiet island of Chek Lap Kok off Lantau Island (Dwyer, 1994) was but one of a range of environmental issues to affect the colony in its last days.

There is, for example, the new expressway to the airport, the massive scale of the airport itself, as well as the town of 160,000 people at Tung Chung beside the airport. The latter will provide labour for the airport itself as well as related economic activities, which will in itself have a considerable environmental impact, and its population will in turn be affected by the noise and other pollution from the airport itself. The defenders of the airport point out that 300–500,000 people will be saved from the current noise pollution at Kai Tak, which is to be closed, and also point out that they saved a rare species of frog that lived on the site.

As regards energy use in this city of lights, Britain signed an agreement to take electricity from the nearby Daya Bay nuclear power plant in China in 1986, despite over 1 million signatures protesting against this potentially dangerous site near the border. More protests followed, post-Chernobyl, about a second nuclear plant at the same site, all to no avail, and the plant was opened in 1994 by Li Peng, then Premier of the PRC. The danger of an accident remains, however, and causes continuing concern.

Conservation is yet another issue. Due to high land prices and development pressures, many of Hong Kong's attractive old buildings, albeit associated with British colonialism, have already been replaced by high-rise structures. Further, there is great pressure on green space, especially on Hong Kong Island and in Kowloon – even high up on Victoria Peak on Hong Kong Island, many expensive houses and small blocks of flats for the wealthy are to be found. The famous bird sanctuary at Mai Po marsh in the New Territories is under threat from development associated with the expansion of Shenzhen just across the border.

Such pressures are unlikely to diminish now that the Hong Kong Special Administrative Region (HKSAR) has been established; in the PRC itself the environmental movement in China is weak, there is a general lack of urban planners, and economic activities take precedence. In 1998, a 'red tide' killed fish in Hong Kong's fish farms while millions of chickens had to be slaughtered in early 1999 to prevent a major flu epidemic. In October 1999, growing environmental pressures led the Chief Executive, Tung Chee-Hwa, to announce strategies to clean up the environment and cultivate talent in an effort to make Hong Kong a world-class city. Mr. Tung announced a $30 billion package to improve air and water quality to 'make Hong Kong a green model for Asia' (Yeung, 1999). He said that if the HKSAR was to achieve a status similar to London or New York, its education system had to keep pace with the times and lure the best brains from the mainland and elsewhere. His address was entitled 'Quality People Quality Home', and in it he warned of the dangers of pollution, both to Hong Kong's image, but also people's health. Plans were announced to cut pollution from thousands of diesel vehicles, phase out diesel-powered taxis and minibuses by 2006, control idling engines and increase the use of non-polluting transport. Interestingly, Mr. Tung also announced an agreement with Guangdong Province to curb pollution. A total of $1.4 billion has been earmarked for grants to owners to help them switch to the use of a cleaner fuel – liquefied petroleum gas and penalties 'for smoky vehicles' would be increased to $1000.

That there is a need for urgent action few deny. In March 2000, for example, choking smog shrouded the territory for days on end as pollution rose to a record level in the central business district. Secretary for the Environment Lily Yam renewed the government's anti-pollution pledge after it emerged that a European organization had called off a business conference in Hong Kong because of the poor air (*South China Morning Post*, 15 April 2000). The business community has often complained that the pollution, caused mainly by vehicle exhausts, discourages foreign investment, and tourist organisations say tourism is also suffering. 'I certainly regard clean air as my top priority . . . and the government is fully behind me in this effort', she assured a gathering of the Australian Chamber of Commerce in Hong Kong (ibid.).

This is a serious economic issue. In its annual Outlook Report for 2000, the Manila-based Asian Development Bank warned that the HKSAR had to improve its environment 'or risk hampering its economic recovery and reputation as an international city'. The report blamed ineffective border enforcement and low fines for contributing to regional diesel pollution in the Pearl River delta. And it highlighted Hong Kong's need to improve its environment if it wanted to become the leading hi-tech hub as the mainland competes to become a port of entry for the region (Ehrlich, 27 April 2000).

In a speech to the same chamber of commerce meeting just referred to, Mr Tung promised that, by 2005, total emissions of respirable particulates emitted from vehicles would be cut by 80 per cent. Nitrogen oxide emissions should be reduced by 30 per cent. 'When we meet these initial targets, our air quality will compare favourably with that of major cities in developed countries, such as New York and London'. The programmes would cost $30 billion in the next ten years. 'If we are reluctant to pay the cost today, we will have to pay more in future when the pollution problem gets worse. There can be no greater folly than this.' He also noted that to cultivate the best talent, an education system that focused on 'cultivation, not elimination' was needed, and its development would be a long-term process. His pro-democracy opponents responded that the speech, subtitled 'Positioning Hong Kong for the Twenty-first Century', was hollow and lacked measures to help the working person or boost competitiveness. As Trade Unionist Lee Cheuk-yan said: 'When he says improving the quality of life in Hong Kong, it sounds like he's renovating the first-class cabin for affluent professionals. But he doesn't give a damn about those in the third-class cabin.'

Analysts said that investors were disappointed with the lack of measures to boost the economy, but Mr Tung denied he had forgotten the 6.1 per cent of workers without jobs.

> We have done a lot in the past two years. Tourism will create jobs. Loans for small and medium enterprises will create jobs. IT development will create jobs. Environmental industry is also big business in foreign countries. It brings new opportunities. I hope the business sector will respond swiftly. Yes. We haven't done similar things to spend and stimulate the economy. But we have and will continue to formulate long-term plans.

It will be interesting indeed to see whether these measures bear fruit, enabling Hong Kong to become, not just a world city of the twentieth century but also one of the twenty-first.

Taiwan pays high price for growth

Taiwan, perhaps eventually to become another SAR, also faces a wide range of environmental problems, many of them emanating from urban areas. The environmental costs of Taiwan's unprecedented economic growth have been high, as Williams (1994) for example has shown. He draws parallels with the similar neglect of environmental factors in Japan's industrialisation, and lists environmental pollution into five types:

1   Biologically active wastes: for example, less than 1 per cent of human excrement receives primary sewage treatment (less than 3 per cent for Taipei), and Hepatitis B is a danger. Many rivers and water bodies suffer from eutrophication.
2   Inert and semi-inert substances: for example, garbage growth was 4 per cent p.a. from 1975–85, and the problem of waste disposal is greatest

for urban areas. There is a lack of landfill sites in such an over-crowded island, and techniques for incineration or separation are costly.

3   Hazardous wastes: are on the increase and a disaster seems likely. 'The general consensus on the island is that problems from hazardous wastes will escalate rapidly within the next decade or two as storage containers disintegrate and the latency periods for human health problems are exceeded' (ibid., p. 242).

4   Atmospheric emissions: 'air pollution' is highly visible, but figures for dust pollution, for example, are disputed. Measures such as catalytic converters are available, however, to reduce the problem.

5   Noise: very high in Taipei, for example, with honking of horns and modern urban life generally. It can be reduced, however.

Williams also refers to such problems as 'eye pollution' which is not a reference to conjunctivitis but to the aesthetic assault on the eye via ugly buildings, billboards and the like, as well as to concerns in more rural areas for loss of land for agriculture (peak production was 922,000 ha in 1977; this was 870,000 ha by the late 1980s and predicted to be 860,000 ha by 2000 (ibid., p. 243)), impending shortage of fresh water due to demographic and lifestyle pressures, wildlife destruction and the need for preserving wild lands. Environmental degradation has become high, in part due to lack of concern by government, by industry as well as by people themselves. However, an environmental movement has developed, as Chen too makes clear (1994). Growing protests took place from 1986, when the environment had become first in a regular poll of people's concerns. The government was forced to upgrade the Bureau of Environmental Protection into a cabinet-level Environmental Protection Agency (EPA) in 1987.

The EPA has been given a hugely enhanced budget (in 1990 it was 20 times what it was in 1987) and was expected to spend a total of $35 billion by the year 2000 (ibid., p. 264). One worry which Chen had, however, is that the EPA will concentrate on the more visible problems of waste disposal, water and air pollution to the potential neglect of more urgent problems such as soil despoliation, toxic wastes and noise pollution.

As an update to Williams's and Chen's concerns, examination of Chapter 13 on environmental protection in the Republic of China's Yearbook, via its website (gio.gov.tw, see references for full details) shows that government policy is aiming at a wide range of issues, including air quality and water resources, but also such concerns as noise pollution and conservation. Air pollution is reckoned to be 'one of the most serious problems in Taiwan, chiefly because of the heavy traffic and high concentration of industrial plants on the island' (Chapter 13, p. 1). The ROC cites data of three vehicles for every five people, 4.02 registered factories and 425 motor vehicles to every square kilometre, that '95 per cent of the air pollution in Taipei', the largest city, is caused by vehicle exhausts, while the Pollution Standard Index (PSI) readings in Taiwan 'are still about three times higher than those recorded in the United States and Europe' (ibid.). To combat this, the Taiwan Air Quality Monitoring Network was set up in 1993, reaching 72 automatic air quality monitoring stations plus two mobile monitoring vans, one air quality laboratory and five remote work stations in 1998. An air pollution control (APC) fee of US$0.006 per litre was imposed on high-grade diesel fuel and on leaded gasoline, plus other taxes on exhausted nitrogen oxides and sulphur oxides, and although the rates seem low, they still generated nearly US$135 million in revenue in fiscal year 1998 to pay for a range of measures, including conversion of motorcar engines to run on liquefied petroleum gas (LPG), a clean fuel (ibid).

In similar fashion:

> Many of Taiwan's rivers and coastal waters have been seriously polluted. Urban communities are major culprits, mainly because of the island's failure to develop its sewage system. Most industrial, agricultural, and residential wastewater drains directly into rivers, seriously polluting the water downstream.
>
> (Ibid., p. 3)

The outcome is that even in Taipei, where construction of the sewage system by the ROC was begun in 1972, 'only 240,000 households (33.22 per cent) had been connected by April 1998' and Taiwan generally has built less than 5 per cent of the total sewage system (ibid.). The number of water quality sampling stations has expanded considerably, to 356 for rivers and 50 for ocean by 1998, and the resultant data shows that 33.8 per cent of primary and secondary rivers 'are polluted to various degrees', the Peikiang being the worst.

Measures are being taken to combat these appalling problems. These include fines of US$2000–20,000 for water pollution, a pollution treatment programme for river basins from 1995, and US$1.44 billion to manage pollution control on the Kaoping and Tungkang rivers. As regards reservoirs these too have great problems, but a 1996 EPA report showed some improvement, with eight heavily polluted reservoirs changing category from 'eutrophic' to 'dystrophic'. However, 'The development of industrial zones, golf courses and real estate presents yet another challenge. Mountain deforestation has severely damaged watersheds' (ibid.). Even this official report notes that 'There seems to be no easy remedy for such up-stream pollution beyond spending more money on down-stream water clean-up or finding new water resources'.

Related to environmental quality is the general problem of energy use. Like Japan, Taiwan has few energy resources, apart from the ingenuity and hard work of its people, so the island is 'about 90 per cent dependent on imported sources of energy' (Williams, 1994, op. cit., p. 251). After the oil shock of 1973–4, the country switched increasingly towards nuclear power, with the first nuclear reactors being opened in the late 1970s. By the mid-1980s, nuclear energy was 18 per cent of total energy use, and a few years later it reached an amazing 40 per cent, 'a level reached by only six nations in the world' (Copper, 1996, p. 138). The growth of the environmental movement, Chernobyl, worries about nuclear waste' and legal complications have meant a rapid reduction since, to 13 per cent of total supply by the mid 1990s. Nonetheless, despite criticisms, 'a fourth nuclear plant is under construction and will go online in the year 2000' (ibid.). The energy situation has still to be fully resolved, therefore, and will continue as an issue, as in other countries, for some time to come.

Analysis: common themes and contrasts

These different parts of 'Greater China' have important roles to play in the environmental future of 'China Proper', as well as in the much-discussed regional question, which has received attention in the last decade. The rich and under-utilised natural resources of Tibet offer a tempting prize for a PRC government struggling to balance its drive for economic development with longer-term sustainable growth. In contrast, Hong Kong and Taiwan offer models for the urban future of the PRC, utilising human resources of knowledge, expertise and entrepreneurial skills to build a potentially high-quality urban environment. In developing an analytical framework to make sense of the differences and similarities between these areas, the following factors seem important.

Han chauvinism. The interaction between these areas and the Han is and will be, fundamental. The legacy of the 'Middle Kingdom' coupled with the rise of the CCP leads to a belief that the Han are, and should be, the dominant force in the development of Greater China. This is expressed in two broad ways: first, via the writ of a PRC central government that believes it knows best which direction China should take; second, via the territorial expansion of the Han people themselves into the regions that surround the Chinese heartland. In Tibet, both of these combine to lead to the intense clash of culture, ideology and belief between a previously feudal society and the PRC's visions of a strong and modern China. In Taiwan the clash takes a different form, but it must be remembered that Taiwanese society was fundamentally transformed by the two great waves of immigration from the mainland in the seventeenth and twentieth centuries, and remains on tenterhooks concerning the long-term aims of the PRC for fusion of Taiwan with the mainland. Hong Kong is already fused, but tensions remain due to the combined legacy of British ideas and concepts with a Cantonese mentality, which can offer a marked contrast with the CCP mind-sets of Beijing. How such tensions are played out between the Han and these three locations will influence the type of environment problems and policies that emerge in each, notwithstanding the level of local autonomy that exists.

Environmental preconditions. Of course, preconditions, too, feed in to environmental outcomes. Here, there are two broad contrasts, between the often virgin environment of Tibet and the despoiled environments of Hong Kong and Taiwan. Tibet offers a high-altitude upland environment, including vast areas of permafrost on the Qinghai–Tibet Plateau that may be uniquely fragile. Hong Kong offers a mainly urbanised environment just within the tropics in a monsoon regime, which can lead to 'frequent heavy downpours . . . sometimes in cloudbursts, which bring phenomenal and unwelcome falls. On 19 July 1926 there was a fall of 100 mm in one hour and over 530 mm in 24 hours. In a typhoon the downpour is accompanied by disastrously strong winds' (Tregear, 1980, pp. 341–2). There is a different type of fragility there which is easily forgotten when we view the sleek skyscrapers of central Hong Kong, never mind when the human dimension is added, with the problems of atmospheric pollution and waste disposal discussed earlier. Also as shown previously, Taiwan offers a broadly similar picture to Hong Kong, lying in the middle of the common typhoon paths for instance, and with marked contrasts in some polluted parts to the 'Ilha Formosa' ('Beautiful Island') of the Portuguese. More research needs to be done for each area to detail the exact ways in which each environment is potentially vulnerable to policies of legislators who fail to fully appreciate the fragility of the environment on which they act.

Economic imperatives. Each of the three areas is subject to the imperatives of economic development, imperatives that often seem to take a marked precedence over environmental concerns. With reference to Tibet, for example:

> China has embarked on a vigorous campaign to publicise its version of a thoroughly modern Tibet, under the guidance of Beijing, that will embrace tower blocks and apartment buildings, private enterprise and tourism.
>
> (Gittings, 17 May 2001)

For almost 1300 years after its foundation as the economic, political and cultural centre of Tibet, Lhasa remained almost unchanged, covering an area of 3 sq. km and with a population of around 30,000 – a situation that still prevailed in 1951. In the 50 years since then, the city has expanded to cover 51 sq. km, with a population of 138,000. The Chinese Government has unveiled plans for an eventual population of half a million. No modern transport facilities existed in Lhasa before 1951. Now it has 130 km of modern roads – and daily traffic jams (*China's Tibet*, July 2001). While the Chinese see this as a major achievement, and want to see even more, one still has to consider the environmental cost of such development.

At the other end of the scale, in Hong Kong and Taiwan the economic path has already been dominant for many decades, with each becoming one of the legendary four 'little tigers' or 'dragons' that have proved so influential for other developing countries, including the PRC itself. Today, the drive is towards higher value-added activities than basic manufacturing, as expressed in the 'Quality People Quality Home' slogan discussed above for Hong Kong. The environment is beginning to have a higher priority, but for most citizens the bottom line remains the economy. A survey in March 2001 in Taiwan, for instance, showed that environmental awareness is on the rise to a greater level than ever before, 'but (people) are still inclined to put economic interests first' (*Taipei Times*, 29 March 2001).

In searching for seeds of hope for the future, and notwithstanding the negative elements we have discussed so far, we conclude by noting two factors that offer a potential positive dimension to environmental concerns, namely political change and alternative environmental models.

Political change. Recent changes in Taiwan show that political change can occur, with positive environmental results, as two examples show. Taiwan elected new president, Chen Shui-bian of the Democratic Progressive Party (DPP) in March 2000, sweeping the Kuomintang (KMT) from power after more than 50 years of control. Chen appointed a retired air force general, Tang Fei, as premier. Tang resigned some months later, ostensibly on grounds of ill health and exhaustion, but more likely, according to Levine (2001), because he supported the completion of the fourth nuclear power station referred to by Copper above, a project that was behind schedule. The power station is opposed by the DPP as well as by environmental activists and at the time of writing now seems unlikely to be completed. Similarly, the new president has halted the Meinung Reservoir project in southern Kaohsiung County, a project similarly opposed by local and overseas environmental activists (Lu, 2001). This is not necessarily the end of the story for either of these controversial projects, but they are indicative that real change is possible via the democratic process, provided that the other constraints discussed previously do not prevent the democratic mandate being exercised.

Alternative environmental models. In similar vein, freeing up the political situation to a range of ideas, especially to alternative environmental models, can present different solutions to existing problems. The leader of a local citizen's group opposed to the Meinung Reservoir for example, suggests that 'There must be a better way to deal with the conflict between environmental protection and economic development. . . . Must we bring harm to Mother Nature?' (Kao, cited in Lu, 2001, op. cit., p. 3). Exploring that theme, water conservation experts suggest that, rather than the water purification proposal suggested by the new government, the real solutions 'are stricter monitoring of industrial waste disposal, limiting the expansion of high-polluting industries, relocating the pig farms, and construction of sewage systems' (ibid.). Thus, a multifaceted strategy is suggested to deal with what is a multidimensional problem. Similarly, Wang and Wang, Taipei architects, suggest that the pollution problems of that city must be tackled via a Greening Strategy that involves such features as 'The need for a person or community who will act as "project animateurs", effective community participation, flexibility, partnership, and fostering a sense of "environmental stewardship"' (2001).

These are ideas with which we are familiar in the West; what is important is to ensure they are locally developed and locally driven to meet local needs. And we conclude here that non-Western ideas such as those of Tibetan Buddhism could also provide a useful source of models for better human–environment interaction, should the will be there among the authorities for such ideas to be explored.

10 Environmental policies

Addressing the opening ceremony of the 1999 Asia–Kyushu Regional Exchanges Summit in Nanjing, Jiangsu Province, Vice-Premier Li Lanqing declared the Chinese government was 'devoting itself to environmental protection in the face of increasing pressures from the rising population and rapid economic development'. To reach its goal of sustained economic growth, China was adhering to the following guiding principles and policies:

- Pollution prevention and control to be further intensified, birth control policies must continue to be upheld, and economic structures readjusted to be environmentally friendly
- Prevention of pollution should top all priorities to eliminate new source of pollutants
- Wastage to be reduced to the minimum when using natural resources
- Scientific progress and technological innovations to be used to reduce energy consumption and spread the use of cleaner fuel in the interests of environmental protection
- Government control and market mechanisms to be utilised and environmental legislation further advanced
- Public media and education institutions to be mobilised to enhance public awareness of the urgency of environmental protection (*Xinhua*, 9 November 1999).

In the previous chapters we have largely concentrated on the various factors that have contributed to China's environmental problems – historical, political, economic, and demographic. In discussing these problems in some detail, we have referred at various points to some of the solutions that have been attempted or are about to be tried at the time of writing. In this chapter, it is our intention to look at policy solutions in more detail. This will concentrate on three aspects: what the government is doing; what the public at large is being encouraged to do; and the international co-operation China is receiving.

Environmental protection: government monopoly

By and large, environmental protection has been monopolised by the government. The Environmental Committee, comprising officials from various departments of the State Council, to which the State Environmental Protection Administration (SEPA) is subordinated, oversees the work. The latter is responsible for implementing the various laws and regulations, but also has a wider role through the enterprises it has set up to produce environmental protection equipment. Through its specialist newspapers and magazines, it seeks to get the message through to business and the general public, whose efforts are often channelled through 'social environmental organisations'. This work is partially supplemented by the legislative and propaganda activities of the Environmental and Resource Protection Committee of the National People's Congress (NPC), the country's parliament.

Non-governmental activities have been somewhat constrained, perhaps due to political self-protection and avoidance of suspicious activities. There is a very thin line between what is acceptable, and engaging in any activity which might be construed as challenging the Party or government's right to control every aspect of Chinese life or of criticising a policy or project that is a favourite of the leadership (for example, former Premier and now NPC Chairman Li Peng's close attachment to the Three Gorges Dam project). To the Chinese authorities, an organisation like Greenpeace, for example, could develop into a strong political opposition party based on its environmental activities (as with the green parties in Western Europe), so that it must be closely monitored and its activities constrained.

Qing and Vermeer have suggested other factors that should be considered as follows:

1   There is a weak public sense of the need to protect the environment, particularly rural residents who have a low educational level.
2   There is an absence of fiscal encouragement or rewards. Because China's civil society has not yet formed a large class of the well-off, the state does not give fiscal encouragement to enterprises that support environmental protection activities. Therefore there are not enough secure sources of private funding to support environmental protection activities.
3   Owing to China's cultural traditions and long-term autocratic rule, the people have not been used to public engagement in social welfare activities in accord with their own wishes. They also lack the habits of consultation and co-operation. Therefore, any organisation without strong bonds is likely to break up in a hubbub because of trivial differences (Qing and Vermeer, 1999, pp. 145–6). But, as we will show, individual efforts for environmental protection are beginning to emerge.

Considering the amount of criticism that has been voiced in these pages so far, it would be as well to acknowledge, as we did in the opening chapter, that

Table 10.1 Drinking water and sanitation access for selected countries

	Safe drinking water (%)			Sanitation (%)		
	Urban	Rural	Total	Urban	Rural	Total
China	87	68	72	100	81	85
India	86	69	73	44	3	14
Indonesia	35	33	34	79	30	45
Sri Lanka	80	55	60	68	45	50
Japan	100	85	96	100	100	100

Source: The World Bank, *Clear Water, Blue Skies: China's Environment in the New Century* (The World Bank, Washington, DC, 1997), Table 2.2, p. 20.

China's achievements in health and life expectancy over the past four decades have far exceeded what could be expected for a country at its stage of economic development. In large part this can be credited to the central government, which has provided family planning, childhood immunisation, accessible primary health care (particularly for mothers and children), improved nutrition, infectious disease control, better education, and improvements in housing and sanitation. Morbidity and mortality from infectious diseases continue to decline on average in most areas of China, although in remote and poor regions the levels of communicable disease remain much higher than the national averages (*People's Daily*, 18 February 1997).

One reason for the improvement in health has been the close attention paid to clean water and good sanitation, where China compares favourably with other countries in Asia (see Table 10.1). This assumes that residents have access to water for washing and that sewage is removed from the house through outdoor latrines, night-soil collection systems, or flush toilets.

Tables 10.2 and 10.3, meanwhile, show the claimed improvements between 1997 and 1998 in dealing with both industrial and household wastewater and chemical oxygen demand (COD) discharges.

In Table 10.4, drinking water is also one of several benchmark figures that are listed for improvements in the urban living environment in the same two years of 1997/8. Also included are the use of gas, access to tap drinking water and the 'innocuous disposal of household garbage and human waste.

There is no doubt that the urban population increasingly demands these services, as do many in the more advanced rural areas. Rising incomes and improved literacy rates, along with a growing sense of national well being, in the era of economic reform since 1979 has brought about a greater environmental awareness among the overwhelming majority of the Chinese people. As communities have become wealthier and better educated, there has been a push for stronger regulations and enforcement, which government at all levels have had to respond to, willingly or not. The increase in media coverage of pollution accidents – a welcome trend after the Maoist era

Table 10.2 Comparison of industrial/household waste water discharges in 1997 and 1998 (in 100 million tons)

	Total amount	Amount of discharge	% of total discharge	Amount of discharge	% of total amount
1998	395	−201	50.9	194	49.1
1997	416	227	−54.6	189	45.4
Variation	−21	−26	−3.7	5	3.7
Rate of change (%)	−5.0	−11.5		2.6	

Source: State Statistics Bureau, 1999.

Table 10.3 Comparison of industrial/household COD discharges in 1997 and 1998 (in 10,000 tons)

	Total amount	Amount of discharge	% of total discharge	Amount of discharge	% of total amount
1998	1499	806	53.8	693	46.2
1997	1757	1073	61.1	684	38.9
Variation	−258	−267	−7.3	9	7.3
Rate of change (%)	−14.7	−24.9		1.3	

Source: State Statistics Bureau, 1999.

Table 10.4 Level of urban infrastructure facilities

Year	Central heating (1 m sq. m)	Total water supply (100 m cu. m)	Tap water penetration rate (%)	Gas use rate (%)	Per capita greenery area (sq. m)	
1998	861.9	471	96.1	78.9	6.1	58.5
1997	807.5	476	95.2	75.7	5.5	55.4
Variation	54.4	−5	0.9	3.2	0.6	3.1
Rate of change (%)	6.7	−1.1			10.9	

Source: State Statistics Bureau, 1999.

approach of 'the only news is good news' – has also contributed to growing public awareness.

Foul water and dirty air, which might have been accepted in the early decades of the PRC as a sign of growing industrial might, are now seen as threats to the gains in health and longevity already referred to. There is a popular saying in the developed eastern region that 'The house is new, the money is enough, but the water is foul and the life is short' (Report of the Fourth National Conference on Environmental Protection, 1996, p. 32).

The legal framework

From the promulgation of the Environmental Protection Law in 1979, the first of its kind in China, five pollution-control statutes and ten natural resource conservation statutes had been enacted by the end of 1999. A key measure was the Energy Conservation Law, passed on 1 November 1997, which came into force on 1 January 1998. It was seen a likely harbinger of strengthened efforts by the government to prohibit certain new industrial projects that seriously waste energy and employ outmoded technologies. Its scope extended to energy from coal, crude oil, natural gas, electric power, coke, coal gas, thermal power, biomass power, and other energy sources.

Box 10.1 (adapted from a similar box in Cook and Murray [2001]) outlines the key initiatives that have been taken by the government and quasi-governmental bodies. From this, it can be seen that the Chinese government has made progress in the broad field of environmental policy. An Environmental Protection Law was introduced on a trial implementation basis as long ago as 1979, when the economic reform programme was only just beginning, and a fuller, updated law promulgated in 1989. The latter, interestingly, incorporates the concept of 'harmonious development' (sustainable development) described 'as a cycle in which physical or natural rebirth and economic rebirth processes work together' (Liu Tanqi, quoted in Edmonds, 1994, p.232). Many other laws and administrative decrees have also been passed, and the new SEPA is heavily involved in legislation and co-ordination of initiatives in this area.

In 1998, the State Council issued the National Plan for Ecological Environment Construction. In all the official approaches, the principle of 'the polluter pays' is increasingly applied.

Yet, despite the complex system of legislative and policy tools in place, and the network of environmental officials throughout China, compliance with environmental regulations throughout the 1990s was rated in the West as somewhat low, essentially because economic development remains the country's priority at all levels of society.

Shen notes that, despite progress in controlling pollutants and wastes on several fronts, the overall environmental quality continued to deteriorate in the 1990s, because, even though the technology and knowledge were available to address the most serious issues of media-specific and point-source pollution (such as pollution from a factory or sewage plant), government policies and strategies were too conservative and fragmented to make progress.

> Furthermore, environmentally harmful products and services have been left out of Chinese research, education and environmental programmes because such problems involve various industries and multiple government agencies without clearly identifying the responsibility of each organisation.
>
> (Shen, 2001, op. cit.)

Box 10.1 National environmental initiatives

1979 Environmental Protection Law (for trial implementation) promulgated.

1980 China joined the WHO's Global Environmental Monitoring System.

1984 National Environmental Protection Agency (NEPA) established as independent quasi-ministerial body.

1989 Environmental Protection Law (including concept of 'harmonious development') promulgated.

1994 Agenda 21 White Paper includes proposals for Office for the Conservation of Diversity to co-ordinate conservation and research.

1998 NEPA became a full ministry as State Environmental Protection Administration (SEPA); National Plan for Ecological Environment Construction approved and promulgated by the State Council.

2000 Release of the first list of controlled ozone-depleting substances (ODS) – CFC-11, CFC-12, CFC-13, CFC-113, CFC-114, Halon-1211, and Halon-1301 – along with new guidelines to regulate companies that deal hazardous substances; imports of carbon tetrachlorine, an ozone-depleting substance, banned; unlicensed exports and imports of seven other hazardous substances forbidden.

Other laws have been promulgated on Water Pollution, Air Pollution, Pollution by Solid Wastes, Land Administration, Protection of Wild Animals, etc.; plus more than 30 administrative decrees on a range of pollution measures and environmental protection, and more than 600 local laws on the latter.

Nature reserves first established 1956, 136 by end of 1998.

Fifteen biosphere Protection Zones established

China joined World Health Organization's Global Environmental Monitoring System in 1980

Experimental Zones for comprehensive agricultural management to cope with drought, salinisation and wind-blown sand in the North China Plain, loess plateau, Manchuria and other areas set up from 1960s onwards

China Academy of Sciences set up various ecology-related bodies from 1954 onwards.

As part of its efforts to strengthen environmental law enforcement, the government revised its criminal code to punish violations in respect of the environment and resources. This step may provide law enforcement agencies with some power. However, the vagueness of standards in many laws and regulations, coupled with the lack of a comprehensive enforcement regime, has led to a situation where many environmental laws still reflect deals cut between the local environmental protection agencies, SEPA, other ministries, local government bodies, and the polluting enterprises.

> China's environmental laws and regulations have yet to work well due to uncertain administrative authority, lack of necessary funds, inadequate monitoring systems and immature environmental concerns. . . . The effectiveness of formal regulations has been hampered in part by inadequate understanding of how to achieve environmental objectives. . . . policies are fragmented across several governmental agencies with different policy mandates and separate media approaches that focus on specific pollutants. They have failed to address the environment as an interrelated whole.
>
> (Shen, 2001, op. cit.)

Thus, the degree of actual compliance and enforcement depends on the region concerned and the personalities involved. Often, the richer the potential investor, the more strictly environmental policy will be applied (*People's Daily*, op. cit.), perhaps on the grounds that they are too high profile to be ignored, while the small fry can pass virtually unnoticed through the net. However, the elevation of NEPA to ministerial status as SEPA in 1998 has provided it with more leverage and authority in law enforcement, which holds out some promise of overall improvement.

Based on its analysis of the programmes drawn up by prefecture and city governments in acid-rain prone areas – covering about 12 per cent of the country – SEPA predicted a 14 per cent cut in sulphur dioxide emissions nation-wide by the end of 2000 and a 42 per cent reduction by 2010 (based on a calculation that more than 23 million tons of sulphur dioxide a year were discharged into the atmosphere in the late 1990s, causing acid rain to fall in 40 per cent of Chinese territory). According to SEPA, 1,798 sulphur dioxide pollution control projects were due to be conducted before the end of 2010, eliminating 9.77 million tons of sulphur dioxide emissions.

This reduction could be realised through comprehensive prevention and control measures such as modifying the current fuel mix, introducing clean energy and production techniques, and reducing the use of high-sulphur coal, said Qiao Zhiqi, the administration's department director in charge of pollution control (author interview).

Under State Council regulations promulgated in early 1999, the discharges of at least 80 per cent of major industrial enterprises were required to meet state standards by the end of that year. Small coal-fired power plants

Table 10.5 Discharge of main pollutants in 1995 and target for 2000

	1995	*2000*
Fly ash (m tonnes)	17.4	17.5
Sulphur dioxide (m tonnes)	23.7	24.6
Industrial dust (m tonnes)	17.3	17
Industrial fixed waste (m tonnes)	61.7	60
COD (m tonnes)	22.3	22
Oils (tonnes)	84,400	83,100
Cyanide (tonnes)	3,495	3,273
Arsenic (tonnes)	1,446	1,376
Mercury (tonnes)	27	26
Lead (tonnes)	1,700	1,668
Cadmium (tonnes)	285	270
Cr6 (tonnes)	669	618

Source: *China Environmental Yearbook 1996*, p. 115, quoted by Vermeer in Edmonds (1998).

operating for 25 years or more had to be shut down, and unlicensed coalmines and licensed mines producing high-sulphur coal cease activities in the same period.

Production facilities and techniques causing serious air pollution would have to be abandoned. The boilers in all large and medium-sized cities in the control zones were required to switch from coal to clean energy sources such as gas. It is too early to know whether the targets have been met, but past experience suggests there will have been, and will continue to be, many areas which will drag their feet for one reason or another (lack of funds, concern about unemployment and social instability, etc.).

Table 10.5 provides some interesting figures on the discharge of the major industrial pollutants in 1995 and the government's target, set in 1996, for their control by 2000. One interesting element is how modest the goals actually are, in many cases seeking a small reduction, while in others merely hoping to hold the line at a small increase. This would seem to suggest recognition that the task is not an easy one.

Both authors can recall meeting many Chinese officials who adamantly maintain that economic development must come before environmental protection. These latter also disagree about how stringent environmental initiatives need to be to protect the national health while maintaining economic growth. This internal struggle enhances the paradoxical quality of Chinese environmental law, which may at once appear both simple and complex, or lenient and severe.

Economic solutions

Parallel with more stringent legal requirements, there is also a need for various economic solutions to be pursued more rigorously. One key area is an adjustment of the pricing system so that it reflects true environmental costs. Ever since the PRC was founded, energy and water have been priced far lower than actual cost (a way to justify keeping down wages and also keeping down the costs for key state industries), requiring the straitened state budget to bear an increasingly intolerable burden. However, great strides are being made to rectify this situation.

In the final years of the 1990s, the government raised and partly deregulated coal prices so that they largely covered production and delivery costs. In addition, many cities and provinces are trying to increase sewage and water charges to consumers and industries. In Taiyuan, Shanxi Province, for instance, the price bureau announced in 1998 that water prices would quadruple over the ensuing five years in order to recover supply costs. Shanghai increased tap water prices by between 25 and 40 per cent to fund water quality improvement programmes and to make sewage self-financing.

But as the experience of Guangzhou, the powerful southern metropolis, demonstrates, price rises cannot always be used as the way forward. Faced with overwhelming resistance from local People's Congress deputies and the public, the Guangzhou Municipal Price Bureau in March 2000 was forced to drop a plan to increase the price of drinking water by nearly 29 per cent. The Municipal Tap Water Company had decided the previous month to raise the price of one ton of drinking water to about 0.9 yuan, from the present 0.7 yuan beginning from April. Company officials said the price increase was necessary because of rising production costs.

But in an unprecedented move, 13 local congressional deputies opposed to the price increase summoned Price Bureau officials to a special meeting to explain why the public would have to pay more. After meeting with deputies for three hours, bureau deputy director Chen Weizhong and Jiang Jiefeng, chief of the bureau's price department, dropped the plan. They promised that the municipality would improve the quality of the drinking water and review its production costs before another attempt is made to raise the price (Zheng, 2000).

At the same time, environmental protection undoubtedly needs ever increasing funds, which will have to be found from a variety of sources. A World Bank report noted that investing about one per cent of the country's GDP each year, gradually rising to 2.5 per cent over the first quarter of the twenty-first century, if divided roughly equally between air and water investment, would greatly reduce pollution by 2020 (World Bank, 1997a, p. 1). The central government has long attributed the continued deterioration of the environment largely to lack of funding, but this deficiency is slowly being overcome.

More than 80 billion yuan was spent on the environment in 1998, more than one per cent of the country's GDP, according to SEPA minister Xie Zhenhua. This broke down into 24.42 billion yuan on pollution control and 56 billion yuan on infrastructure to improve the environment, mainly for controlling urban air pollution and for preventing water pollution in major rivers and lakes. During the 1991–95 period, annual environmental spending was close to 40 billion yuan, or about 0.8 per cent of GDP, Xie said. State financial institutions like the State Development Bank of China, the Industrial and Commercial Bank of China, and the Construction Bank of China have also listed industrial pollution control as their major loan projects (*China Daily*, 15 March 1999).

Towards the end of 1999, another financing channel was opened up through the creation of a private-sector industrial investment fund specifically to assist the energy conservation and environmental protection industry in China. At least 80 per cent of the available funds would be invested in high-tech projects in the environmental industry, according to a fund spokesman. According to him, new products with a high technological content, great market potential and promise of high returns on investment had begun emerging from China's energy conservation and environmental protection industry, but individual enterprises tended to shrink from investing in an industry that often depends on government budgets for capital and operating funds.

> This is partly due to projects in this industry often requiring a large amount of one-time investments in research and development and for the installation of equipment, but investors then often have to wait a long time before they see returns. But government financing channels are often inefficient and too bureaucratic to meet the environmental sector's pressing needs. Industrial investment funds, with large amounts of capital and professional management, can well counter the negative aspects of the potentially profitable industry

The fund is co-sponsored by the China Energy Conservation Investment Co, CITIC Development Co. Ltd, Haitong Securities Co. Ltd and Liaoning Energy Co. (*China Daily*, 14 November 1999). At the time of writing, it was too early to say if the idea would work, especially in fending off excess government interference, particularly at a local level, seeking to force the fund into favourite pet projects that it might otherwise not consider right.

The demand for investment can only continue to grow. During the Ninth Five-Year Plan period (1996–2000), the government adopted the Trans-Century Green Plan, setting targets for environmental protection for the year 2010. This included stabilising emissions of several pollutants at 1995 levels by the year 2000; treatment of industrial waste water expanding to 70 million metric tons, and the percentage of sulphur dioxide, particulates, untreated sewage, and heavy-metal discharges being treated would be increased from 19 per cent to 25 per cent. NEPA originally estimated the programme's cost at 450 billion yuan, or 1.3 per cent of China's GNP in 1996. In geographical terms, the plan assigned top priority to the east coast and in some inland regions, especially the Hai, Huai, and Liao rivers; the Chao, Dianchi, and Tai lakes; and the south-west China areas of Sichuan and Chongqing, with

Table 10.6 Growth of environmental industry 1987–97 (%)

Period	Enterprises	Employees	Output value	Profits
1987–93	342	880	721	392
1994–7	5.1	−9.6	67	83

Source: Industry data for the various years.

pronounced problems with sulphur dioxide levels and acid rain (NEPA, 1996, p. 12).

To cope, industries and local governments are increasingly looking for new sources of funding, through the 'polluter pays' principle, urban environmental infrastructure funds, and even bank loans. The central government is playing a more supportive role in seeking loans and foreign investment, and implementing economic policies. It intended to increase the proportion of GNP spent on controlling pollution from 0.8 per cent in the mid 1990s to more than 1 per cent at the turn of the century, or approximately 188 billion yuan (ibid., p. 4). Some cities are investing an even higher proportion. For instance, Beijing, Shanghai, and Xiamen decided to allocate up to 3 per cent of their GDP to pollution control while Tianjin set a target of 2 per cent.

The increased spending is creating the right conditions for the development of a viable environmental protection industry. By the end of 1998, 9,090 government organisations and enterprises were involved in the industry, with the number of employees increasing from 200,000 in 1988 to 1.69 million, according to SEPA statistics obtained by the authors. Output in 1998 totalled 52.2 billion yuan, of which about 45 per cent involved the manufacture of environmental protection products, the SEPA data showed. Comprehensive resource utilisation (or recycling of major wastes), another major segment of the industry, generated 39 per cent of total output.

In Table 10.6, we show the growth of the environmental protection industry between 1987 and 1997, and in Table 10.7, we provide the actual performance figures for 1997. The impressive percentage increases in the period 1987–93 can be explained by the fact that industry was growing from virtually nothing; the lower figures for the period 1994–7 would seem to be due primarily to the sector achieving a certain amount of maturity. From Table 10.7, it would seem that the most lucrative sector is environmental services (for example, providing factories with the information on how to clean up their act).

The Tenth Five Year Plan (2000–4) will create a tremendous demand for products to tackle a variety of pollution sources, including water pollution, air pollution, liquid garbage and noise. For example, the plan to cap waste water discharges at below 48 billion tons in 2000, and to treat a considerable part of it, creates massive demand for water treatment technologies and equipment (*Business Weekly*, 3 October 1999).

Table 10.7 Financial performance of key elements of the environmental industry in 1997

Sector	Fixed Assets (bn yuan)	Output value (bn yuan)	% of total	Profit (bn yuan)	% of total
Equipment to handle and treat solid waste		2.766	5.3	0.389	6.7
Noise and vibration control equipment		3.39	6.5	0.308	5.3
Environmental monitoring equipment		1.2	2.3	n/a	n/a
Recycling of waste solids, liquids and gases		20.46	39	1.95	33
Recycling of materials (waste paper, metals, etc.)		12.4	23.8	0.72	12.4
Environmental Protection Services		5.28	12	0.83	14.3
Industry total	72	52.2		5.81	

Source: Industry data analysed by the authors.

To generate more cash for these programs, various cities are re-examining their whole attitude towards waste disposal, as we discussed in Chapter 5. Faced with the fact that an urban resident in China at the end of the 1990s was generating 440 kgs of rubbish a year, with that figure likely to grow eight to ten per cent annually (*Business Weekly*, 19 January 1999, p. 8), the question has arisen over who will pay for the cost of collecting and disposing of the mounting piles of rubbish. In most cities, funding comes from the local government budget, but with limited resources, this often means large amounts of rubbish are not being dealt with, especially in a period when priority in spending is being given to industrial development. As a result, Beijing, where only 65.5 per cent of rubbish was rendered harmless in 1988, started to collect a disposal fee from its residents on 1 September 1999. It was calculated that the estimated 130 million yuan to be collected in 2000 would account for about 30 per cent of the municipal government's planned expenditure on dealing with household waste. But Beijing was no trendsetter, as the cities of Nanjing, Kunming, Zhuhai, Wuhan, Chengdu, Beihai, Guilin, and Suzhou had introduced such fees several years before. At the time of writing, Guangzhou and Shanghai were considering similar fee proposals (ibid.).

Zhang Deshan, the man leading the capital's disposal efforts, noted that enterprises in Nanjing paid a rubbish disposal fee based on the cubic metres or tons of rubbish generated, but Beijing has decided to temporarily exempt the corporate sector from the fee for reasons he did not specify. With the municipal government struggling to cope, the question arises of allowing private enterprise to get involved. According to Zhang, no private firm in Beijing was currently engaged in rubbish disposal. Although the government does not forbid such businesses, they are not involved because 'they cannot see an adequate return on their investment as yet. I don't see private firms getting involved until the economic system is reformed to create the right incentives', the official said (author interview).

In contrast, the Chengdu Bureau of Environment and Health Administration in Sichuan Province has successfully involved private firms in the business. A joint-stock company, with the Chengdu bureau holding 30 per cent of the shares and a private firm the remainder, now handles the bulk of household rubbish disposal. The central industrial city of Wuhan also sees rubbish as a valuable asset. Producing 4,000 tons of garbage a day, 14.4 times the level of the early 1950s, and with its landfill sites expected to be full by 2010, the city signed memorandums of understanding in November 1999 with companies from Canada, the United States and the Netherlands to turn rubbish into fertiliser, gas and electricity. ADF Co. Ltd from Canada planned to invest US$45 million to build a garbage treatment plant, capable of transforming 1,000 tons of rubbish into fertiliser, gas and electricity, the American company Hudson agreed to pour a total of US$140 million into a plant of a similar scale, while an unnamed Dutch company planned to use garbage to produce marsh gas which would then generate power (*China Assets News*, quoted in *Business Weekly*, 14 November 1999).

From July 2001, all residents of Tianjin, permanent of temporary, have had to pay for their household trash to be hauled away. Previously the government had borne the full cost, and lack of funds meant most refuse was hauled away for dumping in suburban areas without effective treatment. The new fee ranges from three to five yuan per person, the former for residents covered by property management arrangements.

France and Spain are two other countries heavily involved in helping the major cities deal with their garbage mountains. In Shanghai, for example, Spain has loaned US$30 million towards the US$86 million cost of the Jiangqiao incineration plant, claimed to be the country's largest. The plant was due to open in 2002, handling 1500 tons of garbage a day, enough to treat all household waste from the adjoining Huangpu and Jing'an districts, as well as some from other parts of the city. Meanwhile, the French government has provided a US$30.17 million low interest loan to help Pudong with the estimated US$81 million cost of a similar incineration plant treating 1000 tons of waste a day. France will also provide most of the key equipment. At the time of writing, Shanghai, it should be noted, buried its household rubbish (more than 10,000 tons a day) after a fermentation process. Most of it was handled by the Laogang Waste Treatment Factory, the city's largest landfill site with a daily treatment capacity of 7500 tons; there are also a few small, somewhat inefficient sites in the outer suburbs now regarded as being in very poor condition. One worry for Shanghai is that rubbish treatment costs will increase by five or six times with the new method of incineration, which is more than the environmental protection department believes it can afford. Collection of rubbish treatment fees from citizens was not considered a viable solution because officials expected it would meet with strong resistance (Zhang, 2000)

But the cost factor can be overcome if garbage is treated as a financial resource. Private enterprise in Beijing, both domestic and foreign might be encouraged by a 1999 survey conducted by Beijing Public Health Bureau showing household garbage was 'typically nothing more than misplaced resources'. According to one official:

> . . . out of approximately 13,000 tons of domestic trash produced each day in Beijing, 40 per cent are waste paper, plastic and metal products that can be recycled. One ton of waste paper can produce 800 kilograms of clean paper, while 600 kilograms of unleaded gasoline or diesel can be extracted from 1 ton of waste plastic products, experts say. In addition, waste glass products and clothes constitute a large part of the refuse thrown away.
>
> (Author interview)

Beijing is surrounded by huge mounds of domestic trash, estimated at 600 to 700 million tons, because, 'due to inadequate awareness of the significance of garbage utilisation and outdated technology, most useful materials found in domestic trash cans end up in landfills or burnt to ashes.' (*Beijing Morning Post*, 26 September 1999). One worthwhile approach is to eliminate the materials that cause the disposal problem – an approach being tried by the municipal government in Shanghai (see Box 10.2).

Return of blue skies

The Beijing municipal government, mindful of the state and Party leadership having to live in the midst of some of the country's worst air pollution, has made firm commitments requiring tough action over the first decade or so of the twenty-first century. In this respect, no one has been bolder than the city's Mayor, Liu Qi, who publicly predicted at the start of 2000 'Beijing's skies will be blue for at least 292 days of this year' (*China Daily*, 11 January 2000). The steps the city is taking to meet the goal are described in Box 10.3.

Elsewhere in China, other cities were also making commitments to action. In **Shanghai**, where residents refer to vehicle exhaust fumes as the 'black dragon', had an estimated 600,000 vehicles and 280,000 motorcycles on its streets every day in 1999. The local government planned to have 30,000 light vehicles licensed since 1998 fitted retro-actively with emission controls by the end of 2000, as well as converting 25,000 taxis, 2,000 light trucks and 30 buses converted to use 'clean fuel', usually LPG. Leaded fuel was banned as

Box 10.2 Shanghai promotes 'green sale'

The Shanghai municipal government has launched a 'green sale' project to cut down on waste like plastic packaging, disposable chopsticks, Styrofoam lunch boxes and used batteries. 'This is the first time the city has introduced the "green sale" concept and I believe Shanghai is taking the lead in the country in this regard', said Xia Bojin, director of Shanghai Municipal Food Office. 'The green sale concept combines environmental protection, the development of enterprises and the improvement of quality of life. When the economy is developed to a certain degree, it's time for the city to have a try.'

Three kinds of non-polluting packaging materials were due to appear in the main commercial outlets and streets in the city during the second half of 2000. Biodegradable plastic, which can naturally decompose after five or six months, recyclable plastic and paper will partly replace the now widely used plastic bags in stores and supermarkets. Limits will be placed on the use of excessive packaging beginning the third quarter of the year. 'Excessive packaging is a very serious problem in the city, especially in goods sold in supermarkets such as health care products with layers of paper, plastic and too much padding', said Xia. Disposable chopsticks and Styrofoam lunch boxes were to be banned in downtown main restaurants and snack stores by the end of the year. Provision of disposable articles used in hotels such as cups, slippers and soaps were also expected to be subject to limits. The city's commercial commission set up counters accepting used batteries in large department stores and supermarket outlets.

'But the green sale project will be no easy task', said Xia. The high cost of producing non-polluting products is the first hard nut to crack. The cost of biodegradable plastic is about 70 to 80 per cent higher than ordinary plastic, which frightens off many enterprises. Apart from financial factors, a lack of awareness among local residents also hampers the implementation of the project. The concept of using non-polluting products is not familiar to most people. A lack of non-polluted resources and raw materials is also a big problem for producers. The city plans to develop a range of non-polluted bean products. Special technology and equipment will be used to help grow and package non-polluted beans, currently very rare in China. 'And it is also impractical to put a complete ban on Styrofoam lunch boxes because there are so many restaurants depending on them to earn money', said Xia (Xu, 2000).

As Xia says, such programmes are not easy. A good example is a nation-wide ban on the production and utilization of foam plastic dishware that was due to go into force at the end of 2000 to minimize so-called 'white pollution'. Foam plastic dishware producers argued that the ban was unnecessary and arbitrary. They sought to have implementation postponed indefinitely to avoid the waste of millions of dollars worth of equipment and the loss of tens of thousands of jobs. Liao Zhengpin, President of the China Plastics Processing Industry Association, declared: 'An outright ban is overkill. The dishware causes pollution, no doubt about it, but this problem can be easily resolved through recycling.' He pointed out that six dishware producers from Beijing and Tianjin had invested more than 6 million yuan to establish an enterprise specializing in such recycling and more than 64 per cent of the dishware in Beijing had been recycled in 1999. White pollution, Liao also argued, came not from the production of such products but the bad public disposal habits. (*China Daily*, 12 March 2000)

long ago as 1997, leading the way for the trend that now covers most major cities (*Shanghai Star*, 15 October 99).

In fact, new standards on exhaust emissions for the entire country came into effect on 1 January 2000, promulgated by the State Quality and Technical Supervision Administration (SQTSA), although they were intended to cover only vehicles manufactured after that date. They applied not only unleaded petrol, but also, for the first time, vehicles powered by diesel, liquefied petroleum gas (LPG) and compressed natural gas (CNG).

Regarding Shanghai's waterways, Ge Huizhen, deputy director of the Shanghai Environmental Protection Bureau, said the focus was on getting more of the 696 industrial enterprises identified as still discharging untreated sewage into rivers to clean up their act, although she did not say how this would be enforced. Ge also revealed the city was spending some 38 billion yuan on water treatment projects, as it recognised 'none of the waterways within its boundaries met State standards for drinking use' (*China Dail*, 11 February 2000).

Shanghai is also striving to earn a central government accolade as a 'garden city', a designation so far given to few (see Box 10.4).

Xi'an, the ancient Chinese capital that is home to the renowned Qin Dynasty (221–207 BC) Terracotta Warriors, has decided the best way to protect its environmental reputation with tourists is to switch the main fuel source from coal to natural gas. 'From 1 November [1999], coal burning will be banned in the downtown area within the city wall built in the Ming Dynasty [1368–1644], the high-tech industrial development zone, the economic and technological development zone and the Qujiang tourism zone', declared Vice Mayor Qiao Zheng. In these areas, cleaner fuel, such as natural gas, liquefied petroleum gas and light oil or electricity, will be used for

Box 10.3 Beijing cleans up

Despite Beijing's progress in checking pollution, SEPA Minister Xie Zhenhua told a national conference on environmental protection in January 2000 that much more needed to be done. The government would finance the building and refurbishment of some polluted water and rubbish disposal plants so that, by the end of the year, 40 per cent of dirty water would be treated before it flowing back into rivers, and 80 per cent of Beijing's rubbish would be rendered harmless (*China Daily*, 11 January 2000). The capital began a six-month campaign against air pollution on National Day, 1 October 1999, targeting vehicle exhaust emissions, coal burning and dust from building work in particular. The municipal government said it would rely heavily on 'economic means' – imposing fines – to punish violators. In the winter, the municipal government deliberately lowered the price of 'clean' low sulphur coal to encourage citizens to voluntarily switch from high sulphur fuel long used for heating, as an interim measure pending connection for most city households to a pipeline bringing in natural gas from the Far West (*South China Morning Post*, 1 October 1999). Although containing a certain amount of hyperbole, the following official account does reflect the current mood of hope for lasting improvement.

> It is autumn again. The fall season was not so enjoyable last year. Many people still remember how, for many days, serious air pollution seemed to pack the whole city into a stifling grey box. Every Beijinger knows that it is the massive pollution control project initiated last December that has brought this beautiful season back to this ancient city.
>
> (*China Daily*, 22 September 1999)

In a report in late 1999, Wang Guangtao, vice-mayor of Beijing, insisted the municipal government had enacted 'forceful measures' to cope with various sources and forms of pollution, as follows:

- Since the winter of 1998, it had issued and enforced 18 emergency measures to reduce air pollution. Some 6,700 boilers and 21,000 stoves using coal as fuel had been renovated to use gas, electricity and other clean energy. Coal burning is now banned from the inner city and will gradually expand to the suburbs.
- At the end of 1998, Beijing registered 1.36 million vehicles, 3.4 times more than ten years earlier, with numbers growing an average

14 per cent a year. Discharge of nitrogen dioxide in 1998 was double that allowed, so the municipal government adopted a policy to prohibit pollution-causing emissions in new vehicles, tightened treatment of vehicles currently in use, eased traffic congestion and expanded public transportation. Starting from 1999, it introduced a new standard for automobile emissions (similar to the standard the European Union began enforcing in 1992). According to the new standard, a vehicle can release at most 3.16 grams of carbon monoxide and 1.13 grams of a nitrogen, oxygen and hydrocarbon mixture over one kilometre.

- In a two-year project launched in April 1998, the city spent 336 million yuan on a clean-up of local rivers and lakes, with a pledge that by mid-2000 residents would see 'sparkling, clear green water not seen since the 1960s' (*China Daily*, 15 October 1999). According to the Beijing Municipal Statistics Bureau, the capital had 4,989 hectares of greenbelts by the end of 1999. To celebrate the 50th anniversary of the PRC's founding, the city authorities demolished 4.55 million sq. m of illegal buildings and turned 1.3 million sq. m into greenbelts; 2.24 million trees were also planted last year. The green coverage rate for the urban area of the Chinese capital was calculated at 35.3 per cent (*People's Daily*, 21 February 2000). Meanwhile, a 100 sq. km forest belt was due for completion by the end of 2002, requiring the relocation of many of the city's three million labourers, primarily rural migrants (*China Daily*, 25 April 2000).

On March 26, 2000, Beijing began yet another campaign to shake of its dirty image. The new six-month project targeted excessive automobile emissions, the greening of undeveloped land to limit the dust in the air and converting more boilers from coal to natural gas. The goal was to have 80 per cent of the year with air quality at level three – 'bearable' – or better (levels four and five being considered 'harmful to the body'), compared to 70 per cent in the previous year. SEPA Minister Xie Zhenhua, addressing a visiting delegation of American journalists in May 2000, declared that the government needed to spend 78 billion yuan to clean up Beijing's air and bring the air pollution index down to Category 2 by the end of 2002. This appeared to be in addition to the 47 billion yuan already allocated for the 1999–2002 period.

Box 10.4 Creating a garden city

Dalian, in Liaoning Province, was once a typically grimy port dominated by old-style state-owned heavy industry. In November 1997,
however, NEPA named it and five other major urban centres as a
'Model City of State Environmental Protection', along with a
designation from the Ministry of Construction as a 'Garden City'. The
six (the other five being Shenzhen, Weihai, Xiamen, Zhuhai and
Zhangjiagang) were deemed to have met the following criteria: capacity
for sustainable development; socio-economic level; environmental
quality; environmental management and public participation in
environmental protection. All had urban green coverage of at least
30 per cent, with Xiamen top with 46.9 per cent (Kou, 1998, p. 12). Of
the six, Shenzhen, Xiamen and Zhuhai had the easiest time achieving
high levels of greenery as they were all 'special economic zones' created
largely on former farmland to test the government's theories of
sustainable economic reform. Dalian (population 1.7 million) had a
struggle, given its industrial past, and the first requirement was to get
rid of its polluting industries. From 1995, 84 factories were given their
marching orders and by 1998, 37 of the worst polluters, including
petrochemical plants and breweries had been moved, freeing up some
300,000 square metres of land for redevelopment (ibid., p. 14)

In 2000, Dalian announced a new programme aimed at boosting the
quality of city life. According to officials, residential quarters would
be grouped around schools, hospitals, supermarkets and parks.
Sculptures, fountains and pavilions will also surround each quarter.
At least 40 per cent of each residential quarter in Dalian – which
celebrated its centennial as a city in 1999 – will be covered with
vegetation, said Sun Zhe, an official of the Municipal Management
Office for Real Estate Development. Local government leaders are
ambitious to build Dalian into a modern international metropolis, with
plans to develop trade, finance, tourism and information services by
2010. The city has taken great strides over the past few years to improve
housing conditions as well as the city's environment, according to the
city's Housing Property Administration. A total of 240,000 households have moved into new apartments since the city stepped up
housing construction in 1993. In 1994, the city built special buildings
for primary and middle school teachers and workers in science-related
areas. The city has also put up 'model buildings' for 'model' workers,
whose exemplary work ethic and behaviour other employees are
encouraged to emulate. Sixty new squares and 70 gardens have been
added (author research).

cooking and heating. The installation of new coal-burning boilers was also banned in the remaining urban and suburban districts of Xi'an, with existing coal-burning boilers gradually replaced as conditions improve. In 1998, the total particle content in the air in Xi'an was 0.428 mg per cu, m or 1.24 times the state standard. In the first nine months of 1999, the air content of sulphur dioxide increased from 0.045 to 0.053 mg per cu. m mainly from burning coal (*China Daily*, 20 October 1999).

Shenyang, in Liaoning Province, which has possessed some of the country's poorest quality air and water for many years, actually dropped out of the World Health Organisation's top ten most polluted cities in the world in 1999 thanks to a determined clean-up programme (nevertheless, the National People's Congress still rates it in the top ten on the mainland). The city government was actually the first in the nation to become a member of, and to sign a letter of memorandum with, the International Council for Local Environmental Initiatives (ICLEI), an international environmental agency oriented towards local governments. Shenyang Environmental Protection Bureau chief Liu Tiesheng said the city had made tremendous efforts to relieve pollution, such as cutting the number of household furnaces and restricting the use of coal in winter. Sewage treatment plants capable of handling 500,000 tons of waste a day were opened in the late 1990s, and the city received a US$10 million World Bank loan for a project to deal with toxic industrial wastes (SCMP, 25 September 1999).

Millions of residents living in the vicinity of **Taihu Lake** valley in central China believe they are on the verge of winning a decade-long battle to bring back 'crystal clear water and blue sky'. The provinces of Jiangsu and Zhejiang adjacent to the valley have invested billions of yuan to set up new wastewater treatment centres while taking tough lines against polluters. During a spot check around the valley in March 1999, the local environmental bureau in Zhejiang Province found three out of ten industries failed to meet ecological requirements, and they were closed down instantly for an internal overhaul to meet state standards, said bureau spokesman Zhao Xiao. A total of 257 enterprises on the Zhejiang side, listed as key polluters, had reduced their pollutant release to less than 1,000 tons daily, the state-set environmental requirements, he reported. In Jiangsu Province, meanwhile, 770 key enterprises formerly known as heavy polluters have met required environmental norms after installing or updating pollution treatment facilities, officials there announced.

With the quality of their drinking water worsening over the years, and their land having to be irrigated with polluted water for decades, local residents reacted positively to a government campaign highlighting a 'hot line' telephone they could call to report industrial polluters. The 2400 sq. km lake formerly gave off a foul odour caused by reckless discharge of phosphoric compounds. It created nutrient overloading that led to abundant growth of algae causing serious pollution. Zhejiang Province, together with the Ministry of Construction, will invest 5 billion yuan to set up wastewater

Box 10.5 Bohai Sea detoxification

Two of north-east China's most important ports – **Dalian**, in Liaoning Province, and **Yantai**, in Shandong Province – are at the centre of efforts to tackle the continuously worsening contamination of the ecosystem in the adjoining Bohai Sea, an important fishing ground now becoming depleted largely due to the dumping of industrial waste and the untreated domestic sewage. The State Environmental Protection Administration (SEPA) says the main pollutants are inorganic nitrogen and phosphorus, chemical oxygen demand and oil. The Bohai Sea borders Liaoning, Hebei and Shandong provinces and Tianjin Municipality. Laizhou, Bohai and Liaodong bays are regarded by SEPA the most seriously contaminated (*Business Weekly*, 9 December 1998).

At present, there are 31 rivers and 51 sewer systems dumping sewage into the Bohai Sea – an estimated 2.8 billion tons a year, accounting for 33 per cent of the national total, according to the Shandong Aquatic Products Research Institute. Discharged solid waste totals 700,000 tons, 50 per cent of the coastal areas' total in this category, while the area of the Bohai Sea is only one sixtieth of the state's total ocean area. The sea is relatively enclosed, with an average depth of only 18 metres, and has a limited ability for self-purification. The ecosystem began to deteriorate at an alarming rate from the early 1990s, especially at the estuaries. To counter this, the provincial government is engaged in a three-year plan to clean up the Xiaoqing River, which used to discharge 500 million tons of sewage into the sea every year. More than 100 factories have been closed down. Of an estimated 837 enterprises allegedly contributing to the contamination, 544 had improved their sewage systems and been approved by local governments by the end of 1998 (*China Daily*, 18 January 1999).

According to a draft long-term program agreed in late 1998 by officials from the region and SEPA, industrial enterprises, ships and oil platforms were required to meet national standards for pollutant discharge by 2000. (In 1998, only 46.5 per cent of enterprises in Tianjin, Liaoning and Shandong achieved these standards.) By 2005, water pollution in the coastal areas and damage to the coastal eco-system would have to brought under control, and the quality of groundwater in coastal cities raised to State standards. By 2010, the coastal ecosystem would have to be improved, the discharge of chemical oxygen demand brought under control, a coastal forest belt built, and some ecological demonstration zones established. All pollution control and environmental conservation projects would have to be completed by 2030. Efforts would first be concentrated on cleaning up coastal areas near to cities and river mouths as well as

Laizhou, Bohai and Liaodong bays. The focus would be on 13 cities around the Bohai Sea: Tianjin; Qinhuangdao, Tangshan and Cangzhou in Hebei Province; Jinzhou, Hulu Island, Panjin, Yingkou and Dalian in Liaoning Province; Binzhou, Dongying, Weifang and Yantai in Shandong Province (*Business Weekly*, op. cit.).

treatment centres in urban areas, but domestic users were warned they would have to pay more for their water to make up for the massive investment, they said (*Xinhua*, 23 April 1999).

Meanwhile, east China's **Anhui Province**, a major grain producer, enacted a law bringing the country's fifth-largest lake under legal protection. The Regulations on Prevention and Control of Pollution in Chaohu Lake were adopted by the Standing Committee of the Anhui Provincial People's Congress to stop the deterioration of the 800 sq. km lake, into which some 280 million tons of industrial wastewater and domestic sewage flowed annually during much of the 1990s. The new law required all manufacturers around the lake to install pollution-prevention facilities and report their waste discharges to government departments on a regular basis, with heavy fines and even imprisonment for false reporting (*Xinhua*, 11 June 1999).

The mountainous city of **Yan'an** on the loess plateau in north-west China's Shaanxi Province, home to the Communist Party for much of its struggle for supremacy in the civil war in the 1930s and 1940s, used to be engulfed by thick smoke as local residents used coal-fuelled stoves for cooking and heating. The situation became worse in winter when a multitude of chimneys gushed out heavy black smoke, sometimes bringing visibility down to only a few metres. Local statistics showed that that the urban district of Yan'an had 50,000 stoves consuming 40,000 tons of coal annually in the early 1990s. Most of the coal was bituminous, discharging 3,000 tons of sulphur dioxide and 6,900 tons of smog a year. Zhang Wei, a young woman working in the city's tourism bureau, recalled: 'Coal dust used to stain my collars and cover the white chickens with a thick coat of dust'. A drive by the local government to persuade citizens to switch to low-sulphur and low-ash coal, plus greater use of LPG in vehicles, and several dust control projects (the surrounding loess country is prone to severe dust storms) is beginning to have some effect. With the air quality improving, several large hospitals in Yan'an reported an unusually high ratio of vacant beds in the 1998–9 winter, as the number of people suffering from respiratory diseases decreased (*China Daily*, 1 February 1999).

In **Sichuan Province**, meanwhile hotels and restaurants around the 2256-year-old Dujiang Dam were being demolished to protect the ancient site by cleaning up the environment. Dujiang Dam (Dujiangyan), 40 kilometres north-west of the provincial capital Chengdu, was built on the upper reaches of the Minjiang River in 256 BC during the Warring States period (476–221 BC) and is said to be the world's oldest water conservancy project still operating to serve local agriculture. To protect the area, the provincial government began pulling down more than 1,200 tourism-related structures and entertainment facilities, closing polluting plants, moving residents to other places in the province, and planting trees in the surrounding area (*Xinhua*, 16 February 2000).

These brief snapshots from across the country suggest a pattern of growing environmental awareness at the local level, and raise the possibility that with enough public awareness and cash available, the deterioration can be arrested. But small successes in one area are often offset by worsening conditions elsewhere, ensuring that no one becomes complacent.

International pilot projects

China is very aware that it will be difficult to achieve all its goals alone. It has been turning increasingly to the outside world for help, seeking loans from international aid organisations, working closely with foreign think tanks and environmental organisations, and encouraging foreign investors to put money into the sector – with some success, so far, in the water treatment sector, where French firms are extremely active.

Several pilot environmental protection projects involving international co-operation are now underway, which are helping to popularise advanced environmental technologies and management. According to the China Council for International Co-operation and Environmental Development (CCICED), these include utilisation of biomass energy, rehabilitation of grass vegetation and cleaner energy production. The experiences of these projects are essential for the working group of CCICED to propose more practical and target-geared suggestions to the development of the country, according to SEPA minister Xie Zhenhua.

For example, the biomass demonstration stations in north-east China's Jilin Province, one of the country's major grain production areas, are expected to alleviate the crop-stalk problem. For a long time, large amounts of crop stalk left after harvest were usually burned, causing heavy air pollution in the rural areas. The first pilot project, designed for small villages and townships, can generate 1500–2000 cu. m of gas with one ton of stalk, which can supply fuel to 200 households for cooking. A second pilot project can produce high quality gas and heat for densely populated townships and large-scale commercial production by using dry distillation technology. Though still in its early stages, the projects have been welcomed by the local residents, according to Liu Shuying, vice-governor of Jilin Province (*China Daily*, 25 October 1999).

The disposal of crop stalks, in fact, is a major issue in rural environmental degradation. More than 600 million tons usually remain each year after harvests, and, in the past, the stalks were used by farmers as cattle and horse fodder, and fuel. When coal and liquefied gas became the preferred source of fuel, farmers began burning stubble in the fields, leading to heavy air pollution. In September 1997, for example, the large-scale burning of stubble kept three aircraft from landing at Shijiazhuang Airport, Hubei Province, because of low visibility. Carbon monoxide levels were also four to five times higher than normal at that time, local officials reported. Farmers in north China's Hebei Province are now paying to have their crop stalks pulverised, an environmentally-friendly method that helps enrich the farmland, and some farmers around the cities of Beijing and Tianjin have followed suit. The Hebei Agricultural Machinery Administration allocated 10 million yuan in 1999 to help farmers buy pulverising machines at a cost of 2000 yuan in Shijiazhuang. The central government's Agriculture and Finance ministries have also increased allocations for research on alternative uses of stalks.

Another example of international co-operation is the 'Regional Sustainable Development Case Studies for the Pearl River Delta (PRD)', in which Canada's Sustainable Development Research Institute(SDRI) is heavily involved. These are designed to elaborate, test and improve SDRI's conceptual framework for sustainable development, contributing to the Canada–China Framework For Co-operation on the Environment into the Twenty-first Century signed by the two countries in November 1998, under which development of co-operative projects on regional sustainability are a priority. The objective of the PRD project is to test the extent to which ecological, economic and socio-political goals can be reconciled in the Pearl River Delta.

At the time of writing, SDRI was developing a pilot study in Zhongshan City on achieving regional or community sustainability. The aim is to identify major social, institutional, cultural, economic, and political obstacles to sustainable development. The expected outcomes of the project will be efficient and practical development strategies, technologies, processes, options, policies, and/or plans for policy makers to ensure sustainable regional development. Based on the outcomes of the pilot project, the Canadian team will prepare a proposal for a larger Pearl River Delta project and seek funding from both Canadian and Chinese sources.

Sustainable development, this time in rural areas of the heavily polluted Huang-Haui-Hai Plain, is also the central theme of a project being undertaken by the Centre for International Earth Science Information Network (CIESIN) at New York's Columbia University with a heavy emphasis on resource conservation and protection of the environment. The Huang-Huai-Hai Plain, the largest in China, has been formed by the deposition of the lower reaches of the Yellow (Huang) River, the Huai River and the Hai River. It includes the Provinces of Hebei, Henan, Shandong, Jiangsu, and Anhui, and the Municipalities of Beijing and Tianjin. The plain has 18 million ha of arable land, about a fifth of the mainland total, and produces 40 per cent of the nation's wheat, 56 per cent of its cotton, and 25 per cent of its corn and soybean crops. The Chinese government has since 1949 put considerable resources to agricultural development in this area. Although great achievements have been made in developing agricultural production in this area, but there is the constant threat of drought or water-logging, due to the area's monsoon climate (the region's ability to cope with natural calamities is very low), and the shortage of overall water resources is a serious threat to sustainable agriculture development (the Yangtze diversion project discussed in Chapter 7 has a vital bearing on this aspect). In the future, this area will be expected to produce more and more of the nation's agricultural output, placing increasing stress on the environment. This makes the CIESIN project an urgent task. At the time of writing, the emphasis was on working with seven experimental demonstration sites covering different ecological and economic circumstances.

These sites are:

- Two counties in the low plain of the lower reaches of the Yellow River susceptible to drought or waterlogging (in Hebei or Shandong Provinces).
- One county in the plain in front of Yanshan and Taihangshan Mountain (in Hebei Province) in the area of high production and ample natural resources, good communications, and well-developed secondary and tertiary industries. The main task here will be to develop multiple cropping, optimal-yield, high-efficiency and high-quality agriculture and farm-product processing. The production base will be diversified to include vegetables, fruits, meat, milk, and eggs while preventing or easing environment pollution.
- One county in northern Huang-Huai Plain (in Henan Province) with ample natural resources but a relatively under developed economy. The main task will be to raise productivity to develop a commercial base of wheat and cotton and to diversify agriculture to promote the sustainable development of rural economy while protecting the environment.
- Two counties in southern Huang-Huai Plain (in the north of Anhui and Jiangsu Provinces) with good precipitation and temperatures but poor irrigation and water conservation systems. The calcic black soil of the regions has little ability to withstand drought or waterlogging. The project will help to improve the irrigation and drainage conditions, increase productivity and yields of marketable wheat, soybean, rice and other farm products, diversify agriculture practices and crops, and develop township and village enterprises.
- One county along the Bohai Sea, the Yellow Sea, and coastal regions (in Hebei or Shandong Province) with a well-developed economy, prosperous township enterprises and flourishing foreign trade. However, water resources are in short supply and the agricultural development is poor. The main tasks in this area is to develop high-value and export-oriented agriculture of fruits, vegetables, meat, and specialised products,

to diversify processing of grain, livestock, poultry and aquatic products, to reduce water shortages, and to develop water-saving agricultural practices (www.ciesin.org).

Meanwhile, Taiyuan, capital of north China's Shanxi Province, one of the mainland's most polluted cities has been selected as the mainland's first trial area under a UN Environment Programme for promoting clean industrial production – maximum use of resources and energy with a minimum amount of environmental pollution. The city government selected ten enterprises for the first group to implement the project, which mainly involves use of environmentally friendly furnaces, and it was hoped that half the city's large manufacturers would meet clean production standards by 2001 (*Xinhua*, 17 May 1999).

Taiyuan is also the focus of assistance from the Asian Development Bank, which lent US$102 million to help reduce the city's dependence on coal for industrial, commercial and domestic use – blamed for making it the most polluted place in China and the third worst in the world. The bank said the money would be used mainly to build a new coke-oven gas manufacturing plant, but some of the funds would go to construction of two boilers and a district heating transmission network in Datong City, and the establishment of a system to recover methane gas from an existing coalmine (ADB announcement, 6 December 1999).

Individual foreign governments are also involved in growing number of programmes, For example, Japan provided US$150 million to assist three mainland cities become 'model urban areas for environmental development'. The loans cover projects in Dalian to treat gas emissions from plants and programmes will be funded in Chungqingin and Guiyang to promote the use of natural gas and reduce dependence on coals.

China and Japan are also co-operating in harnessing solar power in poor areas of western China. The New Energy and Industrial Technology Development Organization of Japan (NEDO) has been working with the State Development Planning Commission to establish photovoltaic power generation system equipment in the country's remote areas since 1997. The commission puts forward proposals on suitable places for installing the equipment, Japanese experts then carry out further studies and work with the local governments to establish the power sites. In March 2000, for example, solar power equipment was installed in a clinic of north-west China's Xinjiang Uygur Autonomous Region and in a middle school in south-west China's Yunnan Province, bringing to ten the number of sites so far equipped in this way. Two more were scheduled at the time of writing – in Tibet and Qinghai Province (Xu Lan, 2000).

Japan, South Korea and China have begun close co-operation to deal with the international aspects of China's pollution problems, which were referred to at the opening of this book. Meeting in Beijing in February 2000, the environment ministers of the three countries issued a joint communiqué pledging to strengthen their co-operation in environmental protection projects at both regional and international levels. The ministers considered proposals for co-operation to raise public awareness about environmental problems, for projects to prevent fresh-water pollution and land-based marine pollution, and co-operation in the environmental industry. Among the existing projects are the Acid Deposition Monitoring Network in East Asia, and the three countries' joint research project on Long-range Transportation of Air Pollutants. The Beijing meeting was the second, following one in Seoul in 1999.

The Chinese Government and Britain's Department for International Development began co-operating in April 2001 in a £6.77 million project, to be completed by 2005 in Yunnan Province, with the three-pronged aim of managing environmental endeavours in the Jinsha river valley, formulating a provincial environmental protection plan and disbursing aid to impoverished residents in the area to encourage them to follow ecologically sound agricultural practices (for example, no more slash-and-burn).

The European Union, meanwhile, became involved from June 2000 in a comprehensive environmental project with Liaoning Province. One immediate offshoot was the creation of a 'Liaoning Provincial Clean production Centre', weaning local factories of their traditional, heavily polluting production ways into a mode featuring low consumption, low contamination and high yield. A first success was achieved by the Shenyang Hongmei MSG factory, which used to discharge tree million cu. m of waste water a year. New production techniques have drastically cut this and helped the company save money on its production costs, according to Tobias Becker, EU co-director of the project (author interview).

Private foreign companies are also involved in their own pilot projects. The Zurich-based power technology company ABB, for example, began a two-year programme in May 1999 to help China cut pollution from electric power generation by seeking viable energy alternatives to coal. ABB is also helping to improve the efficiency of coal-fired industrial boilers. The company quickly ascertained that China's industrial boilers consume nearly one-third of all coal burned in the country, but most waste natural resources and cause huge air pollution. To solve the problem, it began collaborating with scientists at Qinghua University in Beijing to research the reactivity of unburned carbon in fly ash for improving the combustion efficiency of industrial boilers (*Business Weekly*, 27 February 2000).

Individual initiatives

At times, the official media has sought to shift some of the blame for the country's chronic environmental problems onto the shoulders of the public, with stories of individual greed and short sightedness (for example, illegal logging that leads to flooding, construction of unlicensed local factories that ignore waste disposal regulations). At the same time, however, there are also occasional heart-warming 'good news' stories that relate individual efforts to protect or restore the environment.

One such occurred around Dongting, China's second-largest freshwater lake located in Hunan Province, which fell prey to local people's 'man harnessing nature' movement in the mid-1970s. Back then, so the official *China Daily* related, local farmers were passionately reclaiming farming land from the lake by building dykes, simply for the purpose of getting a higher crop yield (this overlooks the fact that the farmers were not necessarily motivated by greed, but were far more likely to have been responding to the demands of Party cadres). Altogether more than 1,800 sq. km of the lake area was transformed into arable land. Unfortunately, with its water surface so shrunken, the lake could no longer hold much water in the rainy seasons and floods became frequent. After devastating flooding in 1998, farmers in the area began to move to other places to give back the land to the lake, and by November 1999, as many as 148,000 farmers had relocated themselves (*China Daily*, 26 January 2000).

A few citizens' environmental protection organisations have emerged. Dai Qing and Vermeer, for example, mention the China Cultural Academy's Green Cultural Branch (known as the 'Friends of Nature'), established in 1994 to carry out mass environmental education, establish and spread 'green culture with Chinese characteristics', support and promote all government initiatives and expose and hopefully stop any developments that run contrary to environmental protection; Global Village Cultural Centre, founded by returned Chinese scholars in 1995, with special emphasis on the relation between women and the environment; and Green Earth Volunteers, founded in 1996 by journalist Wang Yongchen, former director of the Friends of Nature, which encourages green-related individual volunteer activities (Dai Qing and Vermeer, op. cit., pp. 146–50). The two authors also list flourishing environmental protective activities on university campuses (ibid., pp. 150–2). Other groups, promoted by the government but essentially non-government organisations, include the Chinese Society for Environmental Sciences, China Forum of Environmental Journalists, China Environmental Culture Promotion Society, and China Environmental Protection Fund.

Occasionally, the official media has elevated an individual to national stardom for their environment-related activities. These include: Zhou Meien, a middle-aged woman working for the Xuzhou Mining Bureau's educational department, who organised a 'Little Reporters Group' of local school-children to research pollution problems in the town, a concept which has now become a nation-wide phenomenon; retired journalist Tang Xiyang and photographer Xi Zhinong who highlighted the rapid disappearance of golden monkeys in their natural habitat of Yunnan Province, and managed to bring the surviving 200 or so under grade one State protection; and Beijing hotel manager Chang Zhongming and Hebei province businessman Han Yulin who have spent their own money to plant trees and develop forestry nature reserves.

In the annals of forestation, meanwhile, one must not forgot the late Ma Yongshun, posthumously awarded the title of 'Hero of the Forest'. Ma, born in 1913, began to work as a lumberjack in Heilongjiang's Tieli Forest Centre in 1948. For 40 years, from 1959 until his death, he voluntarily planted trees with the help of family members, leaving behind more than 20 ha of forest. According to government speakers at a commemorative ceremony in Beijing's Great Hall of the People shortly after his death, the lumberjack had

> set a good example for millions of loggers by stopping tree-cutting and planting trees to reduce soil and water erosion. Instead of cutting trees, the traditional work of state-owned forest farms, about 600,000 loggers are turning to tree-planting or protecting and managing existing forests. Ma Chunhua, Ma's daughter, said her father vowed to plant 35,000 trees in his remaining years to make up for the 35,000 trees he had cut as a lumberjack since the 1940s.
>
> (*China Daily*, 25 February 2000)

The 'Hero' title, the first of its kind in China, was awarded by the China National Greening Committee, the Ministry of Personnel and the State Forestry Administration.

Personal involvement is very much part of the government's national forestation programme. During a pilot effort in western regions in 2000, for example, the State Forestry Administration (SFA) aimed to plant trees on 343,300 hectares along the upper reaches of the Yangtze River and the upper and middle reaches of the Yellow River, but also hoped to enlist millions of villagers to plant an additional 780,720 ha of forest and grassland.

The SFA, it might be mentioned in passing, has committed 220 million yuan into a massive forestation plan 'aimed at turning barren and arid western regions into green areas', between 2000–10. Further investment would go to developing new scientific methods of forestation, pledged Zhu Lieke, director of SFA's Department of Science and Technology (*China Daily*, 2 May 2000). Zhu said he was confident that the plan featuring scientific design, construction and management would increase the survival rate of trees and grasses in the western region by at least 10 per cent. 'Instead of only keeping 45 to 75 per cent of the planted trees and grasses alive in western China as in previous years, more than 85 per cent of the forest products are expected to be preserved', Zhu said, explaining that this would involve selection of drought-resistant trees, cultivation of good seedlings, water containment in the soil, aerial forestation and weather modification, particularly, artificial rainfall, use of flora-accelerators, and control of insects and forest fires (ibid.). At the same time, the SFA, State Development Planning Commission and Ministry of Finance are cooperating in a 1.9 billion yuan program to turn almost 340,000 ha of farmland back to forest in Yunnan, Sichuan, Guizhou, Hubei provinces, and the municipality of Chongqing, along the upper reaches of the Yangtze River, and Shaanxi,

Gansu, Qinghai, Shanxi, Henan provinces, and the Ningxia, Inner Mongolia, and Xinjiang autonomous regions on the upper reaches of the Yellow River (*Jinrong Shibao*, 30 March 2000).

Personal involvement is also evident in turning back the encroachment of desert along China's borders, and is described in Box 10.6.

Thus, in these small and scattered ways, is the message that everyone in China has a stake in restoring a green and pleasant land being put across. But individual involvement is still relatively sparse, and will probably remain so unless a more open political system emerges, one which does not equate criticism of government policy with a desire to bring down the Communist Party of China and the current government. In the next and last chapter, therefore, we will consider what else might be done.

Box 10.6 Building up a green Great Wall

The former windy and sandy Zhangwu area, where the people were poor and land was desolate, has been transformed into lush green land by a mix of official and personal efforts. Zhangwu County is located on the southern border of Kelqin Desert, in Liaoning Province. Before 1949, the Kelqin Desert moved quickly to the south due to disorderly reclamation and destruction of the forest and brush cover, forming a floating sandy belt in the north of the county. This created an unproductive area stretching 50 kilometres from east to west, and 15 kilometres from north to south. From 1949 onwards, however, efforts at sand fixation by planting pine and poplar trees and grass, have created a forest zone described locally as a Green Great Wall. In recent years this has received help from the Belgian Government and the United Nations Food and Agriculture Organization. The 'Fixing Sand by Bushes' concept pioneered by Zhangwu has now spread through ten provinces and autonomous regions, including Gansu, Ningxia and Qinghai.

Since 1979, a large number of people have begun contracting to work on the sand dunes to return them to productivity. In 1984, Yang Haiqing, a farmer of Xiangxiahe Village, then aged 34, contracted for 1000 mu of dunes and planted 120,000 pine saplings. He nursed the saplings day after day until they became a defensive wall five kilometres long and two metres high, finally becoming a genuine forest populated by cattle and sheep. In 1988, Miao Sheng, a middle-aged electrician of the Bureau of Electricity for Agriculture Use, living in Hanjia Village, contracted 500 mu of sandy land and started planting trees in his spare time. Within ten years, he had planted 3000 poplar trees (author research).

(continued overleaf)

While desert has been expanding by 2460 sq. km annually on a national basis, shrinkage at a rate of 1.62 per cent a year in north-west Shaanxi Province's Yulin Prefecture, is offered a proof that the trend is reversible. The Mu Us Desert, lying across the boundary of the Inner Mongolia Autonomous Region and Shaanxi Province, covers 24,400 sq. km, accounting for 56.7 per cent of the total land area in Yulin Prefecture (population: 1.1 million).

But, according to Director Li of the Yulin Forestry Bureau, after nearly half a century of ceaseless efforts, a shelter belt network has taken shape. By the end of 1996, the prefecture had nearly 1 million ha of woods and grasslands, with its vegetation coverage rate rising from 1.8 per cent in 1949 to 39.8 per cent. Many birds and wild animals that had vanished for many years have reappeared in the region. The number of sandstorm-affected days has declined from 66 to 24, and the amount of sand blown into the Yellow River has been cut by 76 per cent. This achievement was recognized when Yulin hosted the International Symposium on the Development and Utilization of Desert Land in August 1992.

According to Li Xiongwu, Deputy Administrative Commissioner of Yulin Prefecture, the region's desert control programme has undergone three stages: an experimental stage (1950–60), when pilot projects were launched in state-run farms and areas with suitable conditions, the successful experience helping local farmers overcome their fears of the desert and control; large-scale treatment (1961–79), with a batch of villages excelling themselves in comprehensive desert control; and finally rapid progress since the economic reforms begun in 1979). Xiaotanzi Village in Dingbian County is a miniature of the prefecture's desert control effort in the initial stage.

Shi Haiyuan, the Party secretary of Yulin County between 1981–87, introduced new measures for planting trees and treating the sandy land on a contracted basis, closing the desert and prohibiting the use of desert plants to make wicker ware, a popular local industry and a product much in demand. As soon as he assumed office, Shi signed a three-year desert control contract with the cadres of each township, under which those who fulfilled set tasks were rewarded with a salary rise, a bonus and the title of model worker. The county leaders also formulated the policy of 'unified planning, separate treatment by households, ownership of the treated land by those who treated it, and permission of inheritance'. This helped mobilise villagers' initiative to plant trees and treat the sandy land. By the end of the 1980s, Yulin had treated 88,000 ha of desert land, and patches of woodland began to dot the desert.

On a personal level, everyone knows the story of Niu Yuqin, one of the top ten outstanding women in China, and her husband, who contracted 670 ha of sandy wasteland in 1985. In a few years, the couple and their assistants planted several hundred thousand saplings. In 1991, they settled down in the desert by building a three-room house and drilling a well. Through more than a decade's effort, they have built 1200 ha of forests, creating an oasis in the desert. For her remarkable contributions, Niu was awarded the titles of a national model worker and a national outstanding Communist Party member. The Xi'an Film Studio has made her achievements into a film, entitled 'A Tree', which won a national award (from Chen Jiazhen and Chen Sichang, 2000).

11 Whither China? Alternative environmental futures

In this book we have examined a wide range of environmental concerns relevant not just to the present, but also the past and future of the People's Republic of China. We felt that it would be appropriate to round off our analysis by contemplating alternative scenarios for the future, the better to see what would need to be changed in the next few years in order to achieve the most desirable of these contrasting outcomes.

We begin by putting on rose-tinted glasses to describe a rosy future, in which the fears and concerns expressed in this book are largely proved unfounded. We then replace the lenses, first with somewhat darker ones in order to see a more 'gloomy' set of outcomes, before with the darkest ones to peer anxiously at a 'doomsday' scenario. Finally, we examine 'back to the future' alternatives that seek to draw effectively on those elements of China's history that stress ecological harmony, to determine how and to what degree the lessons from that past could be utilised to improve the quality of China's environment and increase its potential to cope with the pressures it faces.

Rosy

China is a vast and often beautiful country. Its people have long been shown to be capable of enduring hard, even harsh conditions, and to emerge the stronger for the experience. A rosy environmental future would in part be based on this sheer survivability of the Chinese people. The concerns of today would be endured; the worst environmental concerns would be overcome in part by the sheer diligence and application of the masses of China. This scenario is utopian (a word which of course carries the suggestion 'is found nowhere', not just the positive description of an ideal state), often put, and this utopia is one in which, for instance, pollution from China's industries is controlled, China's cities are effectively greened, vehicle pollution is drastically curbed, its degraded soils are replenished, and the desertification process is severely curtailed. The emphasis is put upon working with the environment rather than on its exploitation, and the worst elements of this exploitation are eliminated. Environmental experts receive appropriate training and there are sufficient numbers of them so they can work throughout the country, with local people across the land, to develop and apply best practice to the local situation. From this best practice, models are developed to apply to other places, other situations. These models will not be like that of 'Sputnik', the first commune of 1958, which were applied uncritically across the length and breadth of the country. Instead, they must be adapted and applied to meet the needs of the specific place, with the experts listening to the locals and the locals listening to the experts in partnerships that are themselves harmonious, so that the application can be harmonious with the local environment.

As well as looking to the local situation, the local people, in partnership with the experts, will also look nationally and globally. As we showed in the previous chapter, there are many examples of good practice across China. Similarly, across the world are many more exemplars of good ideas and applications, of greening the local environment, of sustainable development, of excellent attempts to put the ideals of Agenda 21 firmly into practice. In this rosy environmental future, people associated with these ecological innovations, whether they are community workers, local government officials, environmental scientists, civil servants, business people or others will be invited to share their ideas and applications with people from diverse communities across the land. Groups will travel throughout China, people from overseas will visit the PRC, and people from localities across China will make study visits around the globe. Many of these visits may be 'virtual', for the Internet will be widely used to keep costs down, and also to ensure that too much globetrotting does not itself endanger the environment. Networking will be a vital element, ensuring that *guanxi* extends across all spatial scales, from the local to the global.

'Partnership' is an easy word to employ, and has become a vogue term in the West. In our rosy outlook, partnerships will not only be meaningful, but also fundamental to environmental progress. Thus the government of the PRC will not issue top–down edicts to provinces, and those in turn to cities, in turn to counties and in turn again to townships and villages. Rather, each level in the administrative system will be in partnership with those above and below it. Ideas, concerns and solutions will flow two-way, not one-way from the centre. The CCP will be enlightened in this regard, eschewing its knee-jerk tendency towards 'democratic centralism', replacing it with a more enlightened 'democratic decentralism' or some combination of the two, such as 'democratic centralism–decentralism'.

The juggernaut of economic development, which as we have seen in previous chapters currently overwhelms the environment, will be trans-formed. It will continue to roll, but environmental impact assessments will be integral to every proposal for new businesses and to every new structure in the built environment above an agreed threshold level. Pollution will be seen as an indicator of business failure, as a cost that threatens business rather than as a means of enhancing profit. Waste, likewise, will become a sign of malpractice, a stigma for any company or enterprise. Employees at every level will be encouraged to point out bad practice, even be praised and rewarded for doing so, and especially for coming up with solutions that reduce or eliminate noxious emissions and residue. Enterprises will look outwards, not just inwards, for models of better practice, working with their local communities, provincial governors and many others to improve current methods.

Participation will be another key element. So too will be dissent and championing of environmental causes. It will be widely recognised that entrenched bureaucracies and hierarchies tend to conservatism and inertia, and that participation and dissidence are essential for maintaining a healthy environment. Friends of the Earth and Greenpeace will be welcomed with open arms, as will dissidents such as Dai Qing, for the realisation will finally have dawned that the environment is silent and needs concerned individuals to speak up for it. 'Environmental whistle blowing' will therefore be encouraged, not discouraged, by the authorities, who will have acknowledged that in a complex modern socio-economic system, the panopticon of government's eyes cannot see everywhere, nor can it focus its attention in ways that fully encompass the vast range of potential environmental malpractice across China.

In this rosy future, at the international level, China will become an enthusiastic participant in a new World Environment Association (WEA), established by the UN to address global concerns about pollution, climatic change and global warming. This will be an association rather than an 'organisation' because of its commitment to sharing and partnership. Such a WEA rightly shares its acronym with the Worker's Education Association of the UK, a fortuitous coincidence in that it avoids the mistakes of other organisations, such as the WTO, by working *with* different countries and their representatives in an iterative process, rather than *on* them. It propels the environment to the forefront of people's consciousness around the world and tirelessly campaigns for biodiversity, best ecological practice and the saving of threatened species. It forms part of the future not just for humanity but for this fragile earth in general for the twenty-first century and beyond. Ideas will be sought from the most obscure and even bizarre source; not just from 'experts', but from gurus, mystics and ordinary lay people in continual interaction of people with people, and of people with the environment across the world. These people and their ideas will create, or recreate, ecological harmony at all scales and banish for all time the evils of acid rain, deforestation, nitrification and eutrophication, the threat of global warming and depletion of the ozone layer.

In China today lie the seeds of some elements of this rosy future, of this dream of a better tomorrow. But to achieve it would take a dramatic, probably impossible, change in the mind-set and attitudes of CPC cadres, government officials, entrepreneurs and developers, as well as of the masses of people across the whole of China. The environment would have to come first, not second, third, or even lower in the order of priorities; people would have to consider carefully their own practices, banishing those that have too high an ecological cost, and forgoing some elements of material life, such as car ownership, until such time as vehicles that have a negligible environmental impact could be produced – made of recyclable materials and powered by hydrogen and solar energy, for example. The cities would have to change drastically to become markedly greener, but rural areas too would have to change, to encourage people to reach symbiosis with the rural environment and thus to reduce the desire to migrate to the cities, where the ecological trampling is greatest. Would this mean economic downturn, and a return to poverty for many? Not, we would suggest, if resources were redirected; employment would be created in cleaning and greening the environment, in treating ecosystems for the worst kinds of degradation, and of researching better methods that have negligible environmental impact. And, of course, the developed world community would have to assist this process by reining in its own rampant consumerism and providing the financial and technical assistance that the PRC would require. Such countries, too, would have to ensure they did not continue to dump their worst-polluting industries in China, as the previous chapter showed they have frequently done in the past.

Gloomy

A more gloomy, and to us more realistic, scenario for the future would be more of the situation found today. SEPA, for example, would continue to publicise the worst examples of environmental malpractice, to fight for its share of resources in order to clean up the environment, and to develop well-meaning regulations, laws and policies in an attempt to keep pace with the ongoing deterioration in ecological conditions witnessed in this book. They would fail, however, for they could not measure up to the scale of the problem. Their regulations, rules and edicts would catch only relatively few culprits, and those the small fry rather than the big sharks, because SEPA's personnel, expertise and power would be insufficient to enable them to deal effectively with the many polluters and others who degrade the environment. The economy, in stark contrast, would continue to be given by far the highest priority, from the local through to the national level, as would other expenditure on defence for instance. Mayors of cities, provincial governors, Party cadres at all levels will speak of 'sustainable development', but essentially they will mean 'continued economic growth' when they use the phrase. Agenda 21 policies will be developed and applied, but these will be top–down rather than bottom–up, and will fail to change the hearts and minds of those, from ordinary citizen through to top leader, who will continue to give priority to material goods and economic expansion, even if the environment degrades further as a result.

For many years to come, one of the main problems will be China's continued huge reliance on fossil fuels, especially coal. Low-sulphur coals will be insufficient in quantity or deemed too costly to extract relative to high-sulphur coals to warrant wholesale replacement of the latter, except in a few high-profile locations, especially Beijing, where fines for using high-sulphur coal will continue to bite, though in an inconsistent and piecemeal fashion. There will continue to be some extraction of sulphur at source, as in some power stations, but the cost of fluid-bed technology, as with other pollution technology, will be regarded as too high to be fully applied across the land. The international community will involve itself in such technological fixes to a certain extent, as at present, but will mainly focus on the commercial possibilities of such environmental improvements, to the detriment of sharing technologies for the good of the environment generally at the global level. Organisations such as the World Bank or Asian Development Bank will fund some models and examples of good practice, and appropriate technological transfer, but this will be at a minimal level compared to the huge scale of the problem.

Urban dwellers will pressure the authorities to take action to ameliorate the worst pollution effects, as in the recent campaigns against pollution from Shougang steelworks in Beijing for instance, but few will be willing to make sacrifices, such as giving up their aspirations to be car-owners, reducing commuting or sorting their own waste into recyclable categories, to enable fundamental progress to be made. For their part, urban authorities will seek to implement a range of environmental regulations, but will not divert resources sufficiently from economic expansion to enable significant progress to be made. A blind eye will continue to be turned towards the big developers who wish to build their huge schemes, and there will be little if any attempt to ensure that buildings are heated via solar power, that air conditioners are replaced by better designs that maximise insulation and warm air flow in winter, and cool air flow in summer, and that lighting is energy-efficient and based as far as possible on natural light rather than electricity.

The urban expansion discussed previously in Chapters 5 and 8 will continue to exert severe pressures on China's environment. The massive drive to expand cities will, for a time, be successful economically, but it will be at enormous environmental cost that will gradually be seen to have an economic impact. Basic sewage facilities, the problems of waste disposal, atmospheric pollution from vehicles and industry, the sheer ecological tramplings of unbridled consumerism, will all contribute to a gloomy environmental future for the many urbanites who flock to these cities, desperate to escape the grind and poverty of rural life and to find their pot of gold at the end of the migration route. Some will indeed find such wealth but most will not, and they will be exposed to respiratory diseases, cancerous substances, and epidemics due to the noxious substances and sheer over-crowding they will experience in these increasingly debilitating urban environments.

As for those who remain rural dwellers, most will continue to cut down too many trees, thus increasing erosion, to dump raw sewage into waterways and to overuse chemical fertilisers, thus adding further to nitrates in water courses. The rural population will increasingly consist of women, children and the elderly as emigration of young and middle-aged men removes them (*China Population Today*, April 1998), and those left behind will find it increasingly difficult to tend to the land in ways that retain soil and water quality and protective cover. In addition, TVEs (Chapter 4) will continue to flout environmental regulations for the sake of a quick profit, and although the polluter, if caught, will indeed pay, the numbers who are caught and made examples of will be relatively small, partly because of the lack of environmental personnel to monitor pollution across the length and breadth of the country, and also because of corruption and resistance by local vested interests. In the north-west, desertification will proceed at a rapid rate, notwithstanding some localised successes in its retardation, as farmers seek to squeeze crops out of fragile soils, to cut down trees to expand areas of cultivation or pasture, and to overgraze in marginal environments. Irrigation will continue to be a problem, with low water quality due to pollution, salinisation and alkalinisation outpacing the time and energy of local people to deal effectively with such issues.

At a wider scale, China's floods will be frequent and large-scale, as deforestation continues despite central edicts, and as processes of climatic change already in train unbalance the hydrological cycle. The floods will co-exist, as now, with the drying up of many waterways, as discussed in Chapter 7, due to overbuilding of reservoirs, lowering of the water table through over extraction of water and other causes. The authorities will become increasingly anxious and stringent in their policy-making but will fail to keep pace with these fundamental contradictory consequences of over-exploitation of China's water resources. The Yangtze dam will be completed, and its hydro-electric energy contribution will be welcome but nonetheless quickly shown to be insufficient given the contemporary demands along the Yangtze basin, especially in the delta region. Meanwhile, some of the worst fears of the dam's critics (see Chapter 6) will begin to be realised as downstream species face extinction, pollution levels build up and as fears grow that the Yangtze faces an environmental disaster of the first magnitude.

The Chinese government will continue to suppress environmental dissidents, accusing them of scaremongering or threatening the security of the state by giving ammunition to its enemies. The clampdown will prove counter-productive, as there is no independent source of grassroots criticism to alert authorities to breaches of environmental regulations. The PRC will also continue to berate the West for not doing more to assist in environmental clean-ups, and will accuse it of hypocrisy in not doing more to reduce pollution levels. And they will be right to a large extent, for the West, for its part, will be mixed in its own environmental policies, with European and North American politicians generally failing to grasp the nettle of atmospheric pollution for fear of losing power via the complaints and action of major industrialists or the powerful oil and road lobby. The impending crisis will indeed require some action, including international co-operation, but it will be a case of too little too late to effectively deal effectively with the consequences of two hundred years of environmental destruction. This future is indeed gloomy, in China and also globally; it is also, unfortunately, highly probable.

Doomsday

At its worst, the doomsday scenario could be triggered in several, most likely interrelated ways. We have seen earlier (Chapter 2) that historically China has suffered periodically from flood and famine, with many millions being affected. The difference today is that there are not only many more Chinese than ever before, at around 1300 million at the turn of the millennium, but they are increasingly concentrated in urban centres, as Chapters 5 and 8 showed. Cities are exciting and dynamic places but they are also peculiarly vulnerable. There is little more poignant than the sight of a city being destroyed, whether by fire, flood, war or other cause. Further, urban dwellers produce little of their own food, and are largely dependent for water and energy on national rather than local resources. Destabilisation of food production and of water or energy provision could quickly engender a crisis. A terrorist attack on the Sanxia dam, an earthquake triggered by the weight of this dam, a nuclear accident or other such disaster could set in chain a sequence of events that led to enormous loss of life and a huge environmental disaster with repercussions across China and beyond.

War would be the most extreme cause of doomsday. War is so severe an outcome that, understandably, the people affected fear most for their own security and safety. But modern war is putting increasing demands on the environment as a whole. The use of nuclear weapons has an obvious impact on the environment, but so too do biological and chemical weapons, and depleted uranium used in conventional weaponry. The consequences of war is an awful one, not just for the people involved but also for the environment generally. Thus 'Agent Orange', a powerful defoliant used by the United States in the Vietnam War, left a horrendous legacy of cancers and genetic mutations; the deadly 'Gulf War syndrome' is thought to be caused by vaccinations against nerve gas toxic wastes were released in large quantities into the Danube basin in the Kosovo conflict, and into the Persian Gulf as the Iraqis sought to burn the Kuwait oilfields; while land mines are known to cause horrific injuries in countries such as Cambodia or Angola. Hopefully, war will not occur; but, although

> east Asia was more peaceful during the 90s than at any time during the
> past 100 years . . . during the past decade those east Asian governments

have been spending substantially on weapons and military modernis-ation. Prosperity has grown but so, it appears, has insecurity.

(Huxley, 2000)

Huxley notes that China's defence budget was $40 billion in 1999, larger than any other country in the region, although Japan's was $39 billion (compared to that of the UK of $35 billion) next was South Korea at $12 billion and then Taiwan at $11 billion. He notes that a key player is, and will continue to be, the United States, which although it has a substantial lead in integrated hi-tech control and command systems, plus sophisticated weaponry, could nonetheless 'in the not too distant future, find the American homeland as well as its forces overseas held hostage by Chinese and [North] Korean missiles' (ibid.). Such a situation could be precipitated by the continual tension between Taiwan and the mainland, or as our friend and colleague Rex Li has analysed it, via balked trade expectations:

> If Chinese decision-makers were to reach a conclusion that the outside world is determined to impede China's economic progress and suppress its re-emergence as a great power, their expected value of war will become greater than the expected value of trade. To guarantee that it can exploit the vast deposits of valuable resources in the South and East China Sea, control important shipping lanes in the areas, and gain strategic advantages over its adversaries, China might contemplate taking the military option.
>
> (Li, 1999, p. 469)

This would in turn lead to a hawkish and determined response from the United States.

But for doomsday to occur does not require war. It could be that a 'tipping point' is reached, in which, for example, as the Caltech scenario discussed in Chapter 5 shows, the pressures in China's burgeoning cities eventually lead to an explosion of conflict based on the disaffected urban masses, whose rising expectations are not met as their quality of life is reduced to minimal levels because resource provision fails to keep pace. Or, it could be that the decision to subsidise the urban population squeezes the farming population unbearably, leading to rural conflict and the breakdown of control across the vast hinterlands of rural China. Alternatively, the break-up of China might occur as a result of regionalist fragmentation. Cook and Li (1996) explore alternative regional scenarios, and Cook and Wang (1998) the implications of regional conflict for foreign direct investment. Such negative scenarios are essentially similar to the *luan* (chaos) that many Chinese continue to dread.

It might be that, rather than human events suddenly triggering an environ-mental disaster, instead the environment just collapses. The continuous pressure of environmental degradation we have illustrated in this book could reach a point of no return – for example, polluted rivers deteriorate beyond the scope of remedial action (a survey recently reported that many Chinese children thought that the colour of water was black), or soils are so badly affected that they do not respond to increased levels of chemical fertiliser, or deforestation precipitates such high levels of soil erosion that siltation completely clogs up many rivers in an extended dry season while the same rivers suffer severe flash floods in the wet season. Such outcomes are already happening, but they all would pass beyond the limits of positive human intervention. On the large scale, desertification could become unstoppable and huge swathes of China would become dry dustbowls, like those in Kansas in the 1930s; other areas in currently prosperous coastal regions could be inundated due to rising sea levels caused by the melting of the polar icecaps. Agricultural production would then plummet, epidemics would become rife and untold misery would affect countless millions of people. Much of the land would become a wasteland, and the remaining islands of fertility area would be fought over, thus adding to environmental instability through the impact of the weapons used to claim such welcome territory.

The CPC would still be in power, having clung on tenaciously through the vicissitudes of change in the early twenty-first century, but its hold on power would become more and more tenuous as it was increasingly blamed for its previous obsession with economic growth. It would react to dissidence and upheaval with greater ruthlessness, adding to, rather than reducing, levels of instability. Eventually it might be overthrown, but by this time environmental conditions would have deteriorated to such an extent that no government could deal effectively with the legacy of neglect. Attention could then increasingly turn to invading neighbouring countries to provide for China's still vast population, though it had been decimated by years of famine and disaster. These other countries would, however, have been affected by their own neglect of their environment, so environmental crisis is now global. Doomsday comes not just to China but elsewhere too. A grim prospect indeed.

Back to the future

It would not be in anyone's interest, except possibly for the most virulent anti-communist, to see such a doomsday scenario come to pass. Apart from the hellish impact on the Chinese people, the knock-on effects of environmental disaster in China would be extreme. The world as a whole would be affected by major noxious emissions from China, the threat of conflict in the Asia-Pacific region, and the global instability that would result from upheaval in such a key player in global affairs. Even the 'gloomy' scenario has a potential deleterious impact on other countries, contributing markedly to the vulnerability of the global climatic system for example, and to processes such as global warming as at present. China's emissions of halon gases, for instance, have recently been heavily implicated in causing the growing holes in the ozone layer over the poles.

The world's environment is already threatened and some believe that the threshold to environmental disaster may already have been crossed, that remedial action is already too late. But on the assumption that doom-mongers are mistaken, and that something can still be done to avert environmental catastrophe in China and elsewhere, we have decided to end this book by considering a 'back to the future' scenario, to determine what features of China's historical practices might best be brought forward into the present and the future in order to facilitate realisation of the 'rosy' scenario discussed earlier.

Chapter 2 examined some of these ancient practices, not all of which, of course, were environmentally positive. Of those that were, we can begin with the 'Legend of Yu', and similarly the work of Li Ping and his son, of Wang Jin, and in Ming times, Pan Xiushun, all of whom followed in this laudable tradition. Here were worthy officials who gave fully of themselves in the perennial struggle to calm and tame the river dragons. Between them, they developed and applied ingenious techniques such as extracting water from the main river channel when necessary, splitting the main channel to ease the flow, ensuring that the river beds were continually dredged, and that the dykes were kept low rather than being allowed to increase in height to the eventual greater endangerment of the surrounding countryside and its people. Yu eventually attained supreme dynastic power due to his labours, while Li Ping and his son ensured that, as we described in that chapter, the plain of Chengdu became perhaps the most fertile, densely populated and most productive rural area in the world. Here are exemplars to Chinese leaders and officials, and to Chinese people in general, of endeavour, intellectual capacity and practical application, worthy of any age. Their work was sustainable, it employed local labour effectively in the winter season, and it was effective, provided local people organised themselves, or were organised, to ensure that these key tasks of maintenance were conducted on an annual basis.

In the recent Maoist era, discussed in Chapter 3, the legacy of heavy industry and some other activities was often negative as far as the environment was concerned. But in the communes the work done to bring better sanitation practices and health care via the 'barefoot doctors' quite rightly gained world renown for its good practice in taking basic knowledge effectively to the local level. This knowledge was partly Western, but as we have seen also partly Chinese, notably acupuncture and the use of traditional herbal remedies. The emphasis in many communes on water conservancy and irrigation helped build up the 'communal capital' that has unfortunately been overexploited since that period. The classic case study of Liu Ling commune discussed in Chapter 3 shows what can be done, in similar fashion to the projects discussed in the previous paragraph, when local people effectively utilise the off-season to improve basic environmental infrastructure of water provision, terracing and afforestation. As regards the latter, the shelterbelt concept, too, is a notable historical example, dating from over 2000 years ago; although its success has often been exaggerated in relation to the ongoing deforestation that now bedevils China, once again the idea is a good one.

In Chapter 2 we also mentioned the notion of the *yin* and *yang*, and how ancient Chinese wisdom desired the balance of the two. Ecological harmony can only occur where there is balance, for too much *yin* can lead to inertia and passivity, to lassitude in the face of the great power of nature, while too much *yang* leads to too much intervention, to attempts to dominate nature and hence to destabilisation of natural processes. The history of the PRC, we suggested, has often witnessed too much of the latter, rather than balancing it out with the power of the *yin*, and we sought to show this in Chapter 6 in our evaluation of the Sanxia Dam project. One does not have to be a mystic to see that what D.H. Lawrence called the 'spirit of place', or what the Chinese call *qi*, can easily be upset by too much human interference in ecosystems, a point that can be made at a range of spatial scales.

China's accelerating urbanisation might well upset the ecological balance, as we have shown. Cities, however, can be adapted to attain greater levels of harmony with the environment, as an increasing literature on sustainable cities shows. Even here there is an historical perspective, with the lengthy record of urbanisation in China providing exemplars relevant to the present and future. In the past, for example, Chinese cities were often in symbiosis with their agricultural hinterlands, offering goods, security and spiritual succour in return for food and labour. Ancient city planning in China was strong in its unified nature, its 'macro-concept' of the role of the city as the regional centre of the countryside, and its ecological consciousness (Shen Yahong, 1992). In contrast, the overemphasis on experience and techniques led to a neglect of theory and hence, eventually, to stagnation. From this analysis, as Cook has noted elsewhere (Cook, 2000), Shen develops 'seven historical insights' which include the need to develop 'a theoretical system of modern urban planning at an advanced world level and with Chinese characteristics' (ibid., p. 73), to draw on the experience of the 'excellent' ancient traditions, and to include ecological theory. The indigenous tradition, therefore, should be revisited and reinterpreted for the needs of the future.

Within cities there is also the need to preserve and conserve ancient structures where possible, or at least to upgrade and adapt these to the needs of today and tomorrow. The low-level *hutongs* of Beijing and their equivalent elsewhere, the ancient parks and buildings, and the narrow lanes of the old towns and cities should be retained, even in the face of seemingly all-powerful developers and those who support them. For these urban features, we suggest, are not just physical legacies of the past; they also have rich symbolic value and give a depth and meaning to modern China. They represent the past in the present, and carry with them a sense of history and of place. They are often havens of tranquillity in the sea of commerce, amid the hustle and bustle of China's cities. But to the developer or CPC official, they may be seen as legacies of feudalism, colonialism (in Hong Kong) or imperialism, or as sites of resistance to change. The building of a new expressway, a modern office block or hotel, the KFC or McDonalds might be held up by what to developers is some mouldy old building whose time has long gone. Although there is now more awareness among planners and urbanists of the need to preserve these ancient features, not the least for tourists, there are still many battles to be fought if these historical remnants are to survive in the urban landscape of the future.

Summing up

Let us now, finally, draw together the various themes of this book and put forward a few proposals for the coming years. We have shown that China's ecological system has been deteriorating because their resources are being utilized beyond their carrying capacity. From the early 1950s, excessive harvesting of wood, destruction of forests and grasslands for use as farmland, reclamation of wetlands, the building of dykes to reclaim land from lakes and seas, and excessive hunting and fishing have led to ecological imbalance, reduced organic content in farmland, aggravated soil erosion, regional expansion of deserts, salinisation of former farmland, near exhaustion of forest and fish resources, and a reduction in the size of natural habitats of endangered species.

Since the 1980s, China has made some efforts to limit logging, to plant trees and grasses, return farmland to forest and prairie, establish natural preserves and protect endangered species. Still, it is still far from attaining its goal of effecting a comprehensive reversal of the worsening of the national ecological system.

At the same time, the growth rate in heavy industry was very high from the 1950s, when the country began to industrialise in a big way, up to the 1980s, resulting in high levels of air and water pollution. In recent years, much effort has gone into industrial restructuring and adjusting the industrial balance to place greater emphasis on light industry and on non-polluting industrial techniques. This has had some success in cleaning up the environment. However, this has been offset by the explosion in road construction and the number of vehicles on these roads – especially as exhaust gas emission levels have been higher than those permitted in many Western countries.

China's environmental management strategies have focused primarily on pollution control technologies – waste reduction and pollutants removal, treatment and disposal. Environmental laws, regulations and programmes have mostly targeted the control of pollution rather than prevention of it. The pollution-control approach has improved environmental quality to a certain extent, but in general it has not only failed to eliminate pollutants but has also transferred them from one medium to another. For example, waste treatment processes may produce a large amount of sludge and residue that must be treated again prior to disposal so as not to create secondary pollution.

There are no quick fixes, and certainly the solution is not a purely financial one. As Edmonds (1998) observes:

> For most of the Chinese bureaucracy, the goals of economic growth and environmental protection are not seen as mutually exclusive: economic growth will eventually pay for environmental clean-up.
>
> (Edmonds, p. 3)

This is the sort of argument that enables local government to go on encouraging local industry to persist in bad practices. And blind faith in a technological fix – what O'Riordan (1981, p. 376) calls 'technocentrism' – is not the answer, either:

> while a growing number of people in China are aware that science and technology have [exacerbated] ecological problems in many cases, most of the leadership and much of society at large appear to believe that technological fixes can be found for most environmental woes.
>
> (Ibid.)

To achieve stable environmental quality, we feel that China must do the following:

- Continue to encourage a fairly high growth rate in light industry, at the expense of heavy industry, and ensure all industrial sectors meet strict environmental standards.
- Policies and organisational systems that distort prices of resources and products should be abandoned and the resource utilisation rate greatly improved. For example, price subsidies for coal and irrigation water should be progressively removed to encourage enterprises and individuals to limit consumption and reduce pollution.
- The environmental protection system should be gradually optimised. For example, there should be a system for the design, construction and operation of pollution prevention and control facilities accompanying major parts of all new, extension and reconstruction projects. Industrial and mining enterprises that discharge waste in excess of the allowed standard could be charged for doing so and the revenue used for pollution control. An environmental impact evaluation system needs to be put in place and given teeth. Before breaking ground, all construction projects should be required to submit a report on how the environmental impact will be controlled; an environmental protection responsibility system should also be adopted.
- Leading government officials at all levels should be held responsible for local environmental quality, and entrepreneurs held responsible for the prevention and control of pollution within their enterprises. A

quantitative check system for overall control of environmental conditions in the cities should be established.

- The market for pollution emission rights could be nurtured with positive results in the short-term, while steps are being taken to clean up all types of pollutant emission. This system is intended to limit emission of pollutants and facilitate the trading of the right to produce certain quantities of pollutants. Trading the 'right to pollute', it is argued in some circles, is an important tool for solving environmental problems. If some pollution is inevitable, this market mechanism can be used to keep that pollution to a minimum.
- The legal and necessary administrative measures needed to protect the environment have to be developed further. A basic legal system is in place; now the stress has to be on legal enforcement, especially in persuading local government to eschew local protectionist interests and follow national policy. Laws in all policy sectors need to be amended to emphasise pollution prevention concepts. All affected ministries need to follow consistent environmental policies.
- Public works intended to better the environment should be actively developed and local authorities given every incentive to adopt them.
- Local people need to be involved wherever possible in raising environmental awareness and taking environmental action. The government needs to create a social and political environment that encourages rather than discourages grassroots involvement. This means, providing guarantees to the ordinary citizen that speaking out and taking action will not lead to retaliation by local vested interests.
- At the other end of the scale, international environmental networks and partnerships must be consolidated to ensure sharing of the best practices from around the globe,

The cost of implementing many of the needed programmes will not be cheap. But it needs to be fully recognised at all levels in China that failing to tackle the country's environmental problems in a totally dedicated manner will put a brake on future economic growth, quite apart from the deleterious social and political implications discussed in this book. It is not a case of 'can China afford to tackle the legacy of decades of environmental damage?' but, rather 'can it afford not to?'

Industrial pollution prevention is necessary to minimise the use of finite resources and reduce wastes discharged into the environment. Some analysts suggest that major improvements could actually be achieved at minimal cost. According to Shen, 'Chinese factories can achieve over 30 per cent reduction in pollution through improved management, without capital investments' (Shen, 2001, op. cit.).

The myriad of industrial pollution prevention possibilities available to China can be divided into four main categories: good housekeeping, materials substitution, manufacturing modification and resource recovery.

This would help companies to start taking control of the process of environmental change in ways that make economic and operating sense, rather than seeing their own processes controlled by tightening regulations and expectations.

There is unlikely to be any ultimate solution of China's environmental problems until the need to work together in solving them is recognised by every Chinese citizen. This will not be easy.

As we were completing this book, sources within SEPA revealed to us that industrial pollution has been increasing and the ecological situation worsening in some parts of China due to the pursuit of 'pure corporate or local economic benefits'. A survey found that in the first half of 2001, about 30 per cent of the closed factories in various regions had resumed illegal production and their old polluting ways, which would be virtually impossible without the connivance of local officials. Water quality in some areas was far worse in the first half of 2001 than at any time in the previous year. Illegal forest destruction and the hunting of protected wild animals continued unabated. Some organisations and local governments used the excuse of developing the economy to cut down trees and occupy forestland, Zhou Shengxian, head of the State Forestry Administration told us.

SEPA Minister Xie Zhenhua summed up the problem succinctly when he declared that: '. . . ecological deterioration continues [because] the endeavour of improvement lags behind the speed of human sabotage' (china.org.cn 6 June 2001).

It is going to be a long, hard battle.

References

Preface

Cheung Chi-Fai (2000), 'Gobi Pollutants Detected in Tung Chung', *South China Morning Post*, 24 January 2000.

China Daily, 'Nation Draws up Specific Plan to Save Threatened Ozone Layer', 28 January 2000.

—— 'Hong Kong Accountable for Pollution', 4 April 2000.

Reuters, 'Scientist Says Asia Threatens Ozone Layer', 24 January 2000.

www.China.org.cn, 'Worsening Environment: A Challenge to Chinese People', 7 June 2001).

Zhao, Bin (1997), 'Consumerism, Confucianism, Communism. Making Sense of China Today', *New Left Review*, March–April.

1 China's environmental crisis: an overview

CAS (1992), *Survival and Development – A Study of China's Long-Term Development*, The National Conditions Investigation Group Under the Chinese Academy of Sciences, Beijing, Science Press.

Chen, Qian (1997), 'Improve the Eco-environment and Rebuild the Beautiful Mountains and Rivers', *China Environment News* (NEPA), 13 September 1997.

China Daily, 'Dust Blinds Capital City', 5 April 2000.

Cook, I.G. and Murray, G. (2000), *China's Third Revolution: Tensions in the Transition to Post-communism*, London, Curzon.

International Labour Organisation (1996), *Economically Active Population, 1950–2010*, Vol. 1, Asia, Geneva, ILO.

Lam, Willy Wo-Lap (1998), 'Water Policy Drenched by Criticism', *South China Morning Post* 12 September 1998.

Li Junfeng *et al.* (1995), *Energy Demand in China: Overview Report, Issues and Options in Greenhouse Gas Emissions Control Sub-Report Number 2*, Washington, DC, World Bank.

Liu Yinglang, 'Environmental Report Warns of Worsening Pollution', *China Daily*, 4 June 1996.

Ma, Josephine, 'Environment Boss Attacked Over Contaminated Water', *South China Morning Post*, 26 January 2000.

Migot, A. (1955), *Tibetan Marches* (English translation by Peter Fleming), London, Rupert Hart-Davis.

Ministry of Public Health (1996), *Selected Edition on Health Statistics of China 1991–1995*, Beijing, State Publishing House.

Murray, G. (1998), *China the Next Superpower. Dilemmas in Change and Continuity*, London, China Library.

NEPA (1997), *1996 Report on the State of the Environment*, Beijing, National Environmental Protection Agency (Chinese language edition).

New York Times, 'China's Dramatic Success in Cutting CO_2 Emissions', 17 June 2001.

Ninth five-Year Plan [1996–2000] for Environmental Protection and Long-Term Targets to 2010, Beijing, Foreign Languages Press.

Shanghai Star, 'Shanghai Faces Serious Water Shortage Due to Pollution', 25 January 2000.

SOB (2001a), '2000 Bulletin on Sea Levels', Beijing, State Oceanographic Bureau, March 2001.

——SOB (2001b), '2000 Bulletin on China's Marine Environment', Beijing, State Oceanographic Bureau, March 2001.

Shen, T.S., 'China Cannot Dismiss Pollution Hazards Haphazardly', Parts 1 and 2, *China Online*, www.ChinaOnline.com, 18/19 January 2001.

United Nations (1997), *Population Division, Urban and Rural Areas 1950–2030* (1996 revision), New York, UN.

Wang Genxu and Cheng Guodong (2000), 'Eco-environmental Changes and Causative Analysis in the Source Regions of the Yangtze and Yellow Rivers, China', *The Environmentalist*, 20, pp. 221–32.

World Bank (1997a), *World Development Indicators 1997*, Washington, DC, The World Bank.

——(1997b), *Clear Water, Blue Skies: China's Environment in the New Century*, Washington, DC, The World Bank.

WRI (1999a), *The Environment and China*, Washington, DC, World Resources Institute.

——(1999b), *Urban Air Pollution Risks to Children. A Global Environmental Health Indicator*, Washington, DC, World Resources Institute.

www.China.org.cn. 'Worsening Environment: A Challenge to Chinese People', 7 June 2001.)

——(Paper) 'Greenhouse Gas Emissions Fall in China', 18 June 2001.

Xinhua, 'China Losing 20 Lakes Each Year, Study Reveals', 12 January 2001.

Zhang Jianguang (1994), 'Environmental Hazards in the Chinese Public's Eyes', *Risk Analysis*, Vol. 14, No. 2.

2 Ancient legacies

Buchanan, K. (1970), *The Transformation of the Chinese Earth: Aspects of the Evaluation of the Chinese Earth from Earliest Times to Mao Tse-Tung*, London, Bell.

Buck, J.L. (1973), 'Land and Agricultural Resources', Chapter 3 in Wu, Y.-L., *China: A Handbook*, Newton Abbot, Devon, David & Charles, pp. 45–70.

Cook, I.G. and Murray, G. (2001), *China's Third Revolution: Tensions in the Transition to Post-communism*, London, Curzon.

Edmonds, R.L. (1994a), *Patterns of China's Lost Harmony: A Survey of the Country's Environmental Degradation and Protection*, London, Routledge.

Fitzgerald, C.P. (1935 (1986)), *China: A Short Cultural History*, London, The Cresset Library.

Hsü, I.C.Y. (1990), *The Rise of Modern China*, fourth edition, New York and Oxford, Oxford University Press.

Lao Zi (1993), *A Taoist Classic: The Book of Lao Zi*, based on Ren Jiyu's 1985 edition, translated by He, G. *et al.*, Beijing: Foreign Languages Press.

Marmé, M. (1993) 'Heaven on Earth: The Rise of Suzhou, 1127–1150', Chapter 1 in Johnson, L.C. (ed.), *Cities of Jiangnan in Late Imperial China*, Albany, State University of New York Press.

Muldavin, J.S.S. (1997), 'Environmental Degradation in Heilongjiang: Policy Reform and Agrarian Dynamics in China's New Hybrid Economy', *Annals of the Association of American Geographers*, Vol. 87, No. 4, pp. 579–613.

Murphey, R. (1980), *The Fading of the Maoist Vision: City and Country in China's Development*, London, Methuen.

Schafer, E.H. (1967), *Ancient China*, New York, Time–Life International.

Spence, J.D. (1990), *The Search for Modern China*, New York and London, W.W. Norton

Tregear, T.R. (1980), *China: A Geographical Survey*, London, Hodder and Stoughton.

Tuan, Y.-F. (1970), *The World's Landscapes: 1 China*, London, Longman.

Wang, K. (1998), *The Classic of the Dao: A New Investigation*, Beijing, Foreign Languages Press.

Wertheim, B. (1975), *Introduction to Chinese History: From Ancient Times to the Revolution of 1912*, Boulder, Colorado, Westview Press.

Wheatley, P. (1971), *The Pivot of the Four Quarters: A Preliminary Enquiry into the Origins and Character of the Ancient Chinese City*, Edinburgh, Edinburgh University Press.

Zhao, S. (1994), *Geography of China: Environment, Resources, Population and Development*, New York, Wiley.

3 Politics in command

Agence France Presse, '26,000 Cases Probed in Logging Ban Blitz', 15 January 1999.

Buchanan, K. (1970), *The Transformation of the Chinese Earth: Aspects of the Evaluation of the Chinese Earth from Earliest Times to Mao Tse-Tung*, London, Bell.

China Chemical Industry News, 'Agricultural Pollution Causes Heavy Losses', 9 June 2001.

CIFOR (1999), Center for International Forestry Research, www.cgiar.org/cifor.html.

Cook, I.G. and Murray, G. (2001), *China's Third Revolution: Tensions in the Transition to Post-communism*, London, Curzon.

Deutsche Presse Argentur, 'Trees Will Disappear in a Decade, Expert Warns', 25 November 1998.

Edmonds, R.L. (1994a), *Patterns of China's Lost Harmony: A Survey of the Country's Environmental Degradation and Protection*, London, Routledge.

FAO (1997), 'China's Country Report on Forestry', *Asia-Pacific Forestry Sector Outlook Study Working Paper Series: Asia-Pacific Forestry Towards 2010*, Working Paper no. APFSOS/WP/14, Bangkok.

Ministry of Forestry (1995), *Forestry Action Plan for China's Agenda 21*, Beijing, Ministry of Forestry.

Freeberne, M. (1971), 'The People's Republic of China', Chapter 5 in East, W.G., Spate, O.H.K. and Fisher, C.A. (eds), *The Changing Map of Asia: A Political Geography*, fifth edition, pp. 341–447, London, Methuen.

Howe, C. (1978), *The Chinese Economy: A Basic Guide*, London, Paul Elek.

Hu Anyang and Wang Yi (1992), *Survival and Development – A Study of China's Long-Term Development*, The National Conditions Investigation Group Under the Chinese Academy of Sciences, Beijing,: Science Press.

Huang, Jikun (2001), 'Farm Pesticides, Rice Production and the Environment', published by the Economy and Environment Program for Southeast Asia (http://www.eepsea.org/).

Leeming, F. (1985), *Rural China Today*, London, Longman.

Li, Choh-Ming (1967), 'Economic Development', in Schurmann, F. and Schell, O. (eds) (1967c), *China Readings 3: Communist China: Revolutionary Reconstruction and International Confrontation 1949 to the Present*, Harmondsworth, Middlesex, Penguin.

Li Yucai (ed.) (1996), *Development Strategies for Forestry Toward the 21ˢᵗ Century*, Beijing, China Forestry Publishing House (in Chinese).

Mao Yu-Shi (1996), *The Economic Cost of Environmental Degradation in China, A Summary*, Occasional Paper for the Project on Environmental Scarcities, State Capacity and Civil Violence, Washington, DC, World Resources Institute.

Muldavin, J.S.S. (1997), '*Environmental Degradation in Heilongjiang: Policy Reform and Agrarian Dynamics in China's New Hybrid Economy*', Annals of the Association of American Geographers, Vol. 87, No. 4, pp. 579–613.

Myrdal, J. (1967), *Report From a Chinese Village*, Harmondsworth, Middlesex, Penguin.

Natural Forest Conservation Action Program (1997), Implementation Plan), Beijing, Ministry of Forestry.

Pan, L. (1997), *The New Chinese Revolution*, London, Hamish Hamilton.

Schurmann, F. and Schell, O. (eds) (1967a), *China Readings 1: Imperial China: The Eighteenth and Nineteenth Centuries*, Harmondsworth, Middlesex, Penguin.

Schurmann, F. and Schell, O. (eds) (1967b), *China Readings 3: Communist China: Revolutionary Reconstruction and International Confrontation 1949 to the Present*, Harmondsworth, Middlesex, Penguin.

Simmons, I.G. (1996), *Changing the Face of the Earth: Culture, Environment, History*, Oxford, Blackwell.

Smil, V. (1993), *China's Environmental Crisis: An Inquiry Into the Limits of National Development*, Armonk, New York and London, M.E.Sharpe.

South China Morning Post, 'Reporter Hurt for Exposing Illegal Logging', 15 October 1999.

UNDP (1996), *China Environment and Sustainable Development Resource Book II: A Compendium of Donor Activities*.

Xue Muqiao (1981), *China's Socialist Economy*, Beijing, Foreign Languages Press.

Xinhua News Agency, 2 September 1998.

4 Market forces unleashed

Associated Press, 'Clean-up Laws Kill Eastman Film Plant' 28 July 1999.

Beijing Review, 17 November 1997.

Business Weekly, 'Small Iron and Steel Plants Face the Ax', 9 January 2000.

Cao Fenzhong (1997), '*Air & Water Quality Pollution Problems in TVEs*', a policy paper prepared for NEPA.

China Daily, 'Yellow River Water Polluted',' 23 January 1999

—— 'Pollution Closures Will Continue', 10 January 2000.

——'Polluted Water Hurts 255', 17 February 2000.

——'Work Hazards a Big Concern', 11 February 2000.

——'Chemicals Suspected of Starting Fatal Blast', 10 March 2000.

China Environmental Yearbook (1996), Beijing, China Environmental Yearbook Press (Chinese language edition).

Cook, I. and Wang J. (1998), *Foreign Direct Investment in China: Patterns, Processes and Prospects in Dynamic Asia: Business Trade and Economic Development in Pacific Asia*, Cook, I., Doel, M., Li, R. and Wang J. (eds), Aldershot, Ashgate.

Evans, R. (1993), *Deng Xiaoping and The Making of Modern China*, London, Penguin Books.

International Environmental Reporter, 'Polluting Industries Move To China', 3 May 1997.

Muldavin, J.S.S. (1996), 'Impact of Reform on Environmental Sustainability in Rural China', *Journal of Contemporary Asia*, Vol. 26, No. 3, pp. 289–321.

——(1997), 'Environmental Degradation in Heilongjiang: Policy Reform and Agrarian Dynamics in China's New Hybrid Economy', *Annals of the Association of American Geographers*, Vol. 87, No. 4, pp. 579–613.

Murray, G. (1993), *The Rampant Dragon*, London, Minerva Press.

——(1996), *China: The Last Great Market*, London, China Library.

——(1998), *China the Next Superpower. Dilemmas in Change and Continuity*, London, China Library.

NEPA (1997), *Report of the State of the Environment 1996*, Beijing, National Environment Protection Agency (Chinese language edition).

Xu Fang *et al.* (1992), 'Economic Analysis and Countermeasures Study of TVEs Pollution's Damage to Human Health', *Journal of Hygiene Research*, Vol. 21, Supplement.

Shanghai Economic News, 4 October 1998

Smil, V. (1996), 'China Shoulders The Cost of Environmental Change', *Environment*, Vol. 39, No. 6.

South China Morning Post, 'Capacity Blitz Closes Cement Plants', 5 February 1999.

—— 'Cement-plant Closures Raise Clean-up Hopes', 9 February 1999.

Wang, James C.F. (1992), *Contemporary Chinese Politics: An Introduction*, New Jersey, Prentice-Hall Inc.

Yang, Dali (1994), 'Reform and the Restructuring of Central-local Relations', in Goodman D. and Segal, G. (eds.) (1994), *China Deconstructs, Politics, Trade and Regionalism*, London, Routledge.

Xinhua, 22 April 1999

5 Demographic and consumerist pressures

Brugger, B. (1977), *Contemporary China*, London, Croom Helm.

Caltech (1999), 'Which World? Scenarios for the 21st Century', http:–www.hf. caltech.edu-whichworld-explore-china.

Chesneaux, J., Le Barbier, F. and Bergere M.-C. (1977), *China From the 1911 Revolution to Liberation*, Hassocks (Sussex), The Harvester Press Ltd (English translation of French original).

China Daily, 'China Looks Closer at Recycling Conundrum', 24 February 1999.

—— 'Birth Control Policy Has Had Good Results', 29 September 1999.

—— 'Spur Growth of Nation's Automobile Industry', 30 October 1999.

Cook, I.G. and Murray, G. (2001), *China's Third Revolution: Tensions in the Transition to Post-communism*, London, Curzon.

Elvin, M. and Skinner, C.W. (1974), *The Chinese City Between Two Worlds*, California, Stanford University Press.

Fang Cai (1997), 'Stare Into The Sky – When Will It Clear?', *China Environmental News*, 21 January 1997.

Gallup (1998), 'Survey of Chinese Consumer Attitudes and Lifestyles', Beijing, The Gallup Organisation.

Kam Wing Chan (1994), *Cities with Invisible Walls: Reinterpreting Urbanisation in Post-1949 Chin*, Oxford, Oxford University Press.

Murray, G. (1993), *The Rampant Dragon*, London, Minerva Press.

——(1998) *China The Next Superpower: Dilemmas in Change and Continuity*, London, China Library.

NEPA (1996), National Environmental Quality Report 1991–5, Beijing, National Environmental Protection Agency.

Nolan, P. (1990), 'China's New Development Path: Towards Capital Markets, Market Socialism or Bureaucratic Muddle?' in Nolan and Dong (eds) (1990), *The Chinese Economy and Its Future*, Cambridge, Polity Press.

People's Daily, 18 October 1999.

Reuters, 'China's Dirtiest City in Uphill Battle', 2 April 1997.

Schurmann, H. (1966), *Ideology and Organisation in Communist China*, Berkeley and Los Angeles, University of California Press.

South China Morning Post, 'Citizens Urged to Clean Up Act', 23 March 1999.

—— 'Planners Thinking Big on Size of Cities', 19 October 1999.

—— 'Beijing Ejects Non-residents', 2 November 1999.

—— 'Cities Clogged by Rubbish Mountains', 20 December 1999.

State Council (1996), 'Environmental Protection in China', Beijing, State Publishing House.

World Bank (1999), 'Chongqing Urban Environment Project', World Bank Project IDCN-PE-49436, September 1999, Washington, DC

WRI (1999), 'Urban Air Pollution Risks to Children. A Global Environmental Health Indicatot', Washington, DC, World Resources Institute.

Xinhua News Agency, 1 May 1998 and 24 October 1999.

—— 'Pipeline to Divert Yangtze River Sewage', 4 February 2000.

6 The Sanxia dam

Agence France Presse (1998), 'Flood Report, Three Gorges Plan "Wrong"', 8 August 1998.

Agence France Presse (2000), 'Thousands Displaced by Three Gorges Dam to Get New Homes on an Island', 18 February 2000.

Associated Press (1999), 'China Uses Foreigners for Dam Project', 31 August 1999.

Becker, J. 'Grand Revival Plan Brings Tired Groans From Sceptics', *South China Morning Post*, 9 March 2000.

China Daily, 'Poor Ecology Takes its Toll', 13 August 1998.

—— 'Yunnan to Ban Logging of Primeval Forests', 2 September 1998.

—— 'Blame Placed on Climate Changes', 28 April 1999.

Chow Chung Yan (1999), 'NPC Chief Demands Faster Three Gorges Resettlement', *South China Morning Post*, 15 October 1999.

CNN Earth Matters (1999), 'China's Three Gorges Dam – Eco-Boon or Cesspool?'

Cook, I.G. and Murray, G. (2000), *China's Third Revolution: Tensions in the Transition to Post-communism*, London, Curzon.

Dai Qing (1998), *The River Dragon Has Come! The Three Gorges Dam and the Fate of China's Yangtze River and its People*, Thibodeau, J.G. and Williams, P.B. (eds) (1998), translated by Yi Ming, New York, M.E.Sharpe.

Douglas, I., Gu Hengyue and He Min (1994), 'Water Resources and Environmental Problems of China's Great Rivers', Chapter 10 in Dwyer, D. (ed.) (1994), *China: The Next Decades*, Harlow, Essex, Longman Scientific and Technical.

Edmonds, R.L. (1994a), *Patterns of China's Lost Harmony: A Survey of the Country's Environmental Degradation and Protection*, London, Routledge.

Fisher, A. (1996), 'China's Three Gorges Dam: Is the "Progress" Worth the Ecological Risk?', *Popular Science*, August.

Freer, R. (2001), 'The Three Gorges Project on the Yangtze River in China', *Civil Engineering*, 144, February 2001, pp. 20–8.

Goudie, A. and Viles, H. (1997), *The Earth Transformed: An Introduction to Human Impacts on the Environment*, Oxford, Blackwell.

Kolb, A. (1971), *East Asia: Geography of a Cultural Region* (tr. C.A.M. Sym), London, Methuen.

Li Rongxia (1998), 'Military–Civilian Joint Efforts Block Flood', *Beijing Review*, Issue 38.

Meade, R.B. (1997), 'Reservoirs and Earthquakes', Chapter 3 in Goudie, A. (ed.) (1997), *The Human Impact Reader: Readings and Case Studies*, Oxford, Blackwell, pp. 33–46.

Pang Bo (1998), 'Dam Needed to Prevent Flood', *Chinafrica*, Vol. 85, January.

South-China Morning Post, 'Farmers Blame Corruption for Annual Flood Emergency', 30 July 1998.

—— 'Mao-Era Abuses "To Blame"', 11 August 1998.

—— 'Sichuan Bans Logging to Curb Soil Erosion', 24 August 1998.

—— 27 August 1998.

—— 'Officials Take Flak for Poor Flood Measures', 28 August 1998.

—— 'Beijing Willing to Sacrifice Logging Industry', 3 September 1998.

—— 'Resettlement Process to be Speeded Up', 15 October 1999.

Simmons, I.G. (1996), *Changing the Face of the Earth: Culture, Environment, History*, Oxford, Blackwell.

Spence, J.D. (1990), *The Search for Modern China*, New York and London, W.W. Norton

Thomas, D.S.G. and Middleton, N.J. (1997), 'Salinization: New Perspectives on a Major Desertification Issue', Chapter 7 in Goudie, A. (ed.) (1997), *The Human Impact Reader: Readings and Case Studies*, Oxford, Blackwell, pp. 72–82.

Vidal, J. (1998), 'Woman Power Halts Work on Disputed Indian Dam', *Guardian*, 13 January 1998.

Walker, H.J. (1997), 'Man's Impact on Shorelines and Near-shore Environments: A Geomorphological Perspective', Chapter 1 in Goudie, A. (ed.) (1997), *The Human Impact Reader: Readings and Case Studies*, Oxford, Blackwell, pp. 4–19.

Wu Bian (1997), 'Damming of the Yangtse', *Beijing Review*, 15 December 1997.

Wu Ming (1998), 'Resettlement Problems of the Three Gorges Dam'.

Xinhua (2000), 'Thousands Moved for Project', 17 February 2000.

Yabuki, S. (1995), *China's New Political Economy: The Giant Awakes* (translated by S.M.Harner), Boulder, Colorado, Westview.

Zhao, Songqiao (1994), *Geography of China: Environment, Resources, Population and Development*, New York, Wiley.

Zhou Bian (1998), 'Hydropower to Cut Greenhouse Gases', *Chinafrica*, Vol. 85, January.

7 Moving the waters

21DNN.com, 5 April 2001.

Brown, L and Halweil, B. (1998), 'China's Water Shortage Could Shake World Food Security', *World Watch*, July–August issue, 1998.

Chen Guidi, 'Warning of the Huai River', *Ming Pao*, 5 September 1996.

China Daily, 'Qinghai Pleads for Aid Against Soil Erosion', 19 April 1999.

—— 'Guangzhou Suffers from Drinking Water Crisis', 3 May 1999.

—— Water Project to Ease City Drought', 22 February 2000.

—— 'Nation Set to Harness River', 1 March 2000.

—— 'Nation Warns of Water Scarcity', 25 March 2000.

—— 'Fish Kills Spark Action by State', 25 April 2000.

—— *China Daily*, 23 May 2000.

—— 'Crops Crippled by 100-day Drought', 7 June 2001.

China Economic Times, 'Parts of Yangtze to be Dry by 2020', 6 April 1998.

China Environment Daily, 27 April 1998.

China Youth Daily, 13 November 1999.

Economic Information Daily, 11 August 1998.

Economy, Elizabeth (1998), *Case Study of China: A Summary*, Washington, DC, Council on Foreign Relations

Ming Pao, 'Grave Situation of China's Rivers', 16 July 1996.

NEPA (1997), *1996 Report on the State of the Environment*, Beijing, National Environmental Protection Agency.

Pan, Lynn (1985), *China's Sorrow: Journeys Around the Yellow River*, London, Century Publishing.

People's Daily, 18 July 1996 and 25 March 2000.

Smil, V. (1996), *Environmental Problems in China: Estimates of Economic Costs*, East–West Center Special Reports, No. 5 (April 1996).

Wen Hui Bao, 13 January 1997.

World Bank (1997), *Clear Water, Blue Skies: China's Environment in the New Century*, Washington, DC, World Bank.

WRI (1999), *China's Health and Environment: Water Scarcity, Water Pollution, and Health*, Washington, DC, World Resources Institute.

Xie Qingtao, Guo Xinan and Ludwig, H.F. (1999), 'The Wanjiazhai Water Transfer Project, China: An Environmentally Integrated Water Transfer System', *The Environmentalist*, 19, pp. 39–60.

Xinhua News Agency, 13 February 1997 and 4 June 1998.

—— 'More Hydropower Stations to Yangtze', 28 February 2000.

Zhang, Weiping *et al.* (eds) (1994), *Twenty Years of China's Environmental Protection Administrative Management*, Beijing, China Environmental Science Press.

Zhao Huanxin (1998), 'Urbanites Advised to Save Water', *China Daily*, 23 February 1998.

8 Ecological tramplings

Agence France Presse, 'Russia to Help China Build Fast-breeder Nuclear Reactor', 2 June 2000.

Arndt, R. L., Carmichael, G. R., Streets, D. G., and Bhatti, N. (1996), 'Sulphur Dioxide Emissions and Sectoral Contributions to Sulphur Deposition in Asia', *Atmospheric Environment*.

Beijing Review, 'Facts & Figures: 536 New Cities in 50 Years', 22 November 1999.

Belkin, H., Zheng, Baoshan, Zhou, Daixing, and Finkelman, R. *'Preliminary Results on the Geochemistry and Mineralogy of Arsenic in Mineralized Coals from Endemic Arsenosis Areas in Guizhou Province, P.R. China'*, published in the *Proceedings of the National Academy of Sciences*, 30 March 2000.

Business Beijing (1999), Issue 41, various articles.

Business Weekly (2000), 'Small Iron and Steel Plants Face the Axe', 9 January 2000.

Carmichael, G. and Arndt, R (1998), 'Deposition of Acidifying Species in Northwest Asia', International Symposium on East-Asia Atmospheric Trace Gases, 17–19 October 1998, Beijing.

Carmichael G.R. and Arndt R.L. (1995). *Long Range Transport and Deposition of Sulphur in Asia*, Report from the World Bank-sponsored project 'Acid Rain and Emission Reduction in Asia' (1995), Washington, DC, World Bank.

CDM (1999), Camp, Dresser and McKee Inc., 'Wastewater Improvements for China', www.cdm.com/divisions/international/project/china.html.

Chen Qiuping (1998), 'Environmental Protection in Action', *Beijing Review*, Vol. 41, No. 36, 7–13 September, pp. 8–12.

China Electric Power Statistical Yearbook, 1991–98, Beijing, State Publishing House.

China Energy Annual Review 1997, Beijing, State Publishing House.

China Statistical Information and Consultancy Centre (1999), *Statistical Communique of the People's Republic of China on the 1998 National Economic and Social Development*, Beijing, Foreign Languages Press.

China Statistical Yearbook 1999, Beijing, China Statistical Publishing House.

Cook, I.G. (2000), 'Urban and Regional Pressures of Development', in Cannon, T. (ed.) (2000), *China: Economic Growth, Population and the Environment*, London, Macmillan.

Cook, I.G. and Murray, G. (2001), *China's Third Revolution: Tensions in the Transition to Post-Communism*, London, Curzon.

Drennen, Thomas E., and Jon D. Erickson (1998), 'Who Will Fuel China?', *Science* Vol. 279, 1483.

Dong Liming (1985), Beijing: The Development of a Socialist Capital', in Sit, V.F.S. (ed.) (1985), *Chinese Cities: The Growth of the Metropolis Since 1949*, Hong Kong, Oxford University Press.

Financial Times, 17 January 2000.

Foell, W.C. *et al.* (1995), 'Energy Use, Emissions, and Air Pollution Reduction Strategies in Asia', *Water, Air, and Soil Pollution*, Vol. 85, pp. 2277–82.

Guo Aibing (1999), 'Curbing SO_2 Air Emissions: Nation Needs Cash Flow To Succeed', *China Daily*, 8 December 1999, p. 2.

Hu Anyang and Wang Yi (1992), *Survival and Development – A Study of China's Long-Term Development*, The National Conditions Investigation Group Under the Chinese Academy of Sciences, Beijing, Science Press.

Huang Wei, 'Water Pollution, Today's Topic', *Beijing Review*, 22 November 1999, pp. 20–1.

—— 'Life and Death in No-Man's Land', *Beijing Review*, 5 July 1999, pp. 16–19.

Kruger (1997), 'Xian City: Bei Shi Qiao Wastewater Treatment Plant', http://www.kruger.dk/xian.html.Li Wen (1999), 'China's Environmental Conditions in 1998', *Beijing Review*, 12 July 1997, pp. 13–18.

Ögütçü, M. (1999), 'China's Energy Future and Global Implications', Chapter 5 in Draguhn, W. and Ash, R. (eds) (1999), *China's Economic Security*, London, Curzon.

O'Neill, M. (2000), 'Notes on China: Capital Games Idea Could Mean Move For Steel Plant', *South China Morning Post*, 10 January 2000.

People's Daily, 'Warning of Water Shortages', 25 March 2000.

Robertson, L., Rodhe, H., and Granat, L. (1995), 'Modelling of Sulphur Deposition in the Southern Asian Region', *Water, Air and Soil Pollution*, Vol. 85, pp. 2337–43.

Sato, J., Satomura, T., Sasaki, H., Muraji, Y. (1996), 'A Coupled Meteorological and Long-Range Transport Model, and its Application to the East Asian Region'; submitted to *Atmospheric Environment* (1995).

Sharma, M., McBean, E. A., and Ghosh, U. (1995), 'Prediction of Atmospheric Sulphate Deposition at Sensitive Receptors in Northern India', *Atmospheric Environment*, Vol. 29, pp. 2157–62.

Shi Hua, 'Matongs' End in Sight as Locals Dump Old Potties', *Shanghai Star*, 27 September 1999.

South China Morning Post, 'Hong Kong Vows To Combat Air Pollution', 15 April 2000.

State Statistical Bureau (1996), *China Statistical Yearbook 1995*, Beijing: China Statistical Publishing House.

State Statistical Bureau (1999), *China Statistical Yearbook 1998*, Beijing, China Statistical Publishing House.

International Energy Outlook (1998), Study on Alternative Energy and Energy Supply Strategies in China 1998, www.eia.doe.gov/oiaf/ieo/98/home.html.

Takle, E. (1998), *'Fuelling and Feeding an Advancing China: A Global Environmental Challenge'*, prepared for Provost's Workshop on China, Iowa State University, 10 April 1998.

Tang Min (1999), 'Tougher Rules Help City Breathe Easier', *Beijing Weekend*, 19–21 November, p. 16.

www.gio.gov.tw/inf/yb97/html/ch13.htm (Environmental Protection, consulted 15 December 1999).

Xiao Chen (1999), 'From Gray to Green: Beijing Business Feels Breath of Fresh Air', *Business Beijing*, Issue 42, November, pp. 12–16.

Zhou Dadi *et al.* (2000), *Developing Countries and Global Climate Change: Electric Power Options in China*, Washington, DC, Pew Centre on Global Climate Change, May 2000.

Zhou Shunwu (1992), *China Provincial Geography*, Beijing, Foreign Languages Press.

9 Pollution on the periphery

Beijing Review, 'Let Tibetan Chamois Share the Earth with Mankind', 5 July 1999, p. 20.

Chen, D.W. (1994), 'The Emergence of an Environmental Consciousness in Taiwan',

Chapter 9 in Rubinstein, M.A. (ed.) (1994), *The Other Taiwan: 1945 to the Present*, Armonk, New York, M.E.Sharpe.

China's Tibet (2000), Beijing, China Intercontinental Press.

Copper, J.F. (1996), *Taiwan: Nation-State or Province?*, Boulder, Colorado, Westview.

Dwyer, D.J. (1994), 'The Hong Kong Airport Controversy', *American Asian Review*, Vol. 12, No. 2, pp. 89–106.

Ehrlich, J. (2000), 'HK Warned to Tackle Pollution or Lose Status', *South China Morning Post*, 27 April 2000.

Gittings, J. (2001), 'Beijing Accelerates Plans for a Modern Tibet', *Guardian*, 17 May 2001.

Hung, Wing-tat (1994), 'The Environment', Chapter 14 in McMillen, D.H. and Man, Si-Wai (eds) (1994), *The Other Hong Kong Report 1994*, Hong Kong, The Chinese University Press.

Jetsun Pema (1997), *Tibet My Story: An Autobiography*, Shaftesbury, Dorset, Element.

Levine, S. (2001), *Brittanica Book of the Year: Events of 2000: Taiwan*, Chicago, Encyclopaedia Britannica, pp. 500–1.

Lu, M. (2001), 'Water Demand vs. Environmental Protection', 9 January 2000, updated 8 February 2001, http://www.gio.gov.tw, consulted 4 June 2001.

Ovchinnikov, V. (1995), *The Road to Shambala*, Beijing, New World Press.

Platt, Kevin (1999), 'Qomolongma Project Offers Way Forward for Tibet', *Christian Science Monitor*, 4 March 1999.

Taipei Times (2001), Environmental Awareness Rising: Survey', 29 March 2001.

Tibet 2000: *Environment and Development Issues*, Environment & Development Desk, Department of Information & International Relations, Central Tibetan Administration, Dharamsala, India.

Tibetan Bulletin (1999), 'Focus on Tibet's Environmental Destruction', Vol. 3, No. 1, January–February 1999.

Tiley Chodag (1988), *Tibet: The Land and the People* (translated by W. Tailing), Beijing, New World Press.

Tregear, T.R. (1980), *China: A Geographical Survey*, London: Hodder and Stoughton.

Wang, J.T.J. and Wang, R.L.H. (2001), 'Toward a Green Taipei Metropolitan', paper presented to the Joint International Planning Conference, Liverpool University/Liverpool John Moores University, Liverpool, March.

Wang Wen (2001), 'Train Will Run on the Roof of the World', *Beijing Review*, 15, 12 April, pp. 16–17.

Williams, J.F. (in collaboration with Ch'ang-yi Chang) (1994), 'Paying the Price of Economic Development in Taiwan: Environmental Degradation', Chapter 8 in Rubinstein, M.A. (ed.) (1994), *The Other Taiwan: 1945 to the Present*, Armonk, New York, M.E.Sharpe.

Yeung, C. (1999), 'Tung's $30 Billion Green Crusade', *South China Morning Post*, 7 October 1999.

Zhang Tianlu (1997), *Population Development in Tibet and Related Issues*, Beijing, Foreign Languages Press.

10 Environmental policies

Agence France Presse, 'Japan To Extend Loans To China For Pollution Control', 5 February 2000).

Beijing Morning Post, 'Beijing Rubbish a Treasure Trove', 26 September 1999.

Beijing Review, 'China's Environmental Conditions in 1998', 22 July 1999.

Bradbury, I. and Kirkby, R. (1995), 'Prospects for Conservation in China', *Ecos*, Vol. 16, No. 3, 4, pp. 64–71.

Business Weekly, 'ABB Lends Helping Hand', 27 February 2000.

—— 'Beijing to Collect Rubbish Disposal Fee', 19 September 1999.

—— 'China Attacks Pollution Points', 3 October 1999.

—— 'Wuhan Gets Foreign Help on Garbage', 14 November 1999.

—— 'Campaign Launched to Clean up Bohai Sea', 9 December 1998.

Chen Jiazhen and Chen Sichang (2000), 'Green Wonder: Desertification Control in Yulin, Shaanxi Province', *Beijing Review*, Issue 19.

China Daily, 18 January 1999.

—— 'Anti-pollution Measures Clean Up Yan'an, Shaanxi', 1 February 1999.

—— 'Measures to Tackle Pollution of Bohai Sea', 8 March 1999.

—— 'More Spending on the Environment', 15 March 1999.

—— 'Grey Skies Turn to Blue', 22 September 1999.

—— 'Pollution Treatment Bears Fruit', 15 October 1999.

—— 'Tourist City Bans Coal Burning', 20 October 1999.

—— 'Pilot Environmental Protection Projects Bear Fruit', 25 October 1999.

—— 'Stubble Pulverised, Not Burned', 7 November 1999.

—— 'Fund Created to Finance Environmental Projects', 14 November 1999.

—— 'Beijing Encourages Citizens, Work Units to Use Clean Fuels', 11 January 2000.

—— 'When Man and Nature Collide', 26 January 2000.

—— 'Shanghai Plans to Curb Pollution in Waterways', 11 February 2000.

—— 'Hero of Forest Remembered for Tree Planting', 25 February 2000.

—— 'Clearer Days Ahead as Beijing Tackles Pollution', 29 February 2000.

—— 'White Pollution Producers' Urge Recycling', 12 March 2000.

—— 'City Plans Work on Massive Forest Belt', 25 April 2000.

—— 'Massive Funds for Green Project', 2 May 2000.

Chinese Academy of Sciences (1995), *Brief Account of the Chinese Academy of Sciences*, Beijing.

Dai Qing and Vermeer, E.B. (1999), 'Do Good Work, But Do not Offend the "Old Communists"', in Draguhn, W. and Ash, R. (eds) (1999), *China's Economic Security*, London, Routledge.

Edmonds, R.L. (1994a), *Patterns of China's Lost Harmony*, London, Routledge.

Information Office of the State Council (1996), *Environmental Protection in China*, Beijing.

Jinrong Shiba, 'China To Turn Western Farms Back Into Forests', 31 March 2000.

Kou Zhengling (1998), 'Garden Cities Spring Up in China', *Beijing Review*, 2 February 1998.

Li Wen (1999), 'China's Environmental Conditions in 1998', *Beijing Review*, 12 July 1999.

Liu Tanqi, in Edmonds, R.L., *Patterns of China's Lost Harmony*, London, Routledge, 1994.

South China Morning Post, 'Fines for Pollution Offences', 1 October 1999.

NEPA (1996), *The National Ninth Five-Year Plan for Environmental Protection and the Long-Term Targets for the Year 2010*, Beijing, National Environmental Protection Agency.

People's Daily, 'Decision on Public Health Reform and Development by the Central Committee of the Chinese Communist Party and the State Council', 18 February 1997.

—— 'Beijing Strives For A "Greener" City' 21 February 2000.
Report of the Fourth National Conference on Environmental Protection, 1996, Beijing, China Environmental Sciences Press.
Shanghai Star, 'Brakes Put on Exhaust Emissions', 15 October 1999.
Shen, T.S., 'China Cannot Dismiss Pollution Hazards Haphazardly, Parts 1 & 2', China Online, www.ChinaOnline.com, 18/19 January 2001.
South China Morning Post, 'Shenyang Seeks UN Aid with Pollution', 25 September 1999.
Vermeer, E.B. (1998), 'Industrial Pollution in China and Remedial Policies', in Edmonds, R.L. (ed.) (1998), *Managing the Chinese Environment*, Oxford, Oxford University Press.
World Bank (1997), *Clear Water, Blue Skies: China's Environment in the New Century*, Washington, DC, World Bank.
——(1997a), *Can the Environment Wait? Priorities for East Asia*, Washington, DC, World Bank.
www.ciesin. org
Xinhua News Agency, 'Environmentalism Takes Centre Stage' 9 November 1999.
—— ' "Clean" Plan Starts in Taiyuan', 17 May 1999.
—— 'Taihu Valley Wins Ecology Battle', 23 April 1999.
—— 'Law Enacted to Save Lake', 11 June 1999.
—— 'Hotels Demolished to Protect Dam', 16 February 2000.
Xu La (2000), 'Japanese Solar Technology Shines in Western China', *China Daily*, 27 February 2000.
Zheng Caixiong (2000), 'Plan to Increase Price of Water Evaporates', *China Daily*, 25 March 2000.

11 Whither China? Alternative environmental futures

China.org.cn. 'Public Awareness Key to Environmental Protection', 6 June 2001.
China Population Today (1998), 'Special Issue on Aging', Vol. 15, No. 2, April.
Cook, I.G. and Li, R. (1996), 'The Rise of Regionalism and the Future of China', Chapter 9 in Cook, I.G., Doel, M., and Li, R. (eds) (1996), *Fragmented Asia: Regional Integration and National Disintegration in Pacific Asia*, Aldershot, Avebury.
Cook, I.G. and Wang, Y. (1998), 'Foreign Direct Investment in China: Patterns, Processes, Prospects', Chapter 6 in Cook, I.G., Doel, M. A., Li, R. and Wang, Y. (eds) (1998), *Dynamic Asia: Business, Trade and Economic Development in Pacific Asia*, Aldershot, Ashgate.
Edmonds, T (ed.) (1998) *Managing the Chinese Environment*, Oxford, Oxford University Press.
Huxley, T. (2000), 'Insecurity Measures', *Guardian*, 31 January 2000.
Li, R. (1999), 'The China Challenge: Theoretical Perspectives and Policy Implications', *Journal of Contemporary China*, 8–22, pp. 443–76.
O'Riordan, T. (1981), *Environmentalism*, London, Psion.
Shen, T.S., 'China Cannot Dismiss Pollution Hazards Haphazardly, Parts 1 & 2', China Online, www.ChinaOnline.com, 18/19 January 2001.
Shen Yahong (1992), 'Enlightenment From the Development of Ancient City Planning in China in the Ancient Times', *China City Planning Review*, Vol. 8, No. 1, pp. 65–75.

Bibliography

Brown, L and Halweil, B. (1998), 'China's Water Shortage Could Shake World Food Security', *World Watch*, July–August issue, 1998.

Brugger, B. (1977), *Contemporary China*, London, Croom Helm.

Buchanan, K. (1970), *The Transformation of the Chinese Earth: Aspects of the Evaluation of the Chinese Earth from Earliest Times to Mao Tse-Tung*, London, Bell.

Buck, J.L. (1973), 'Land and Agricultural Resources', Chapter 3 in Wu, Y.-L., *China: A Handbook*, Newton Abbot, Devon: David & Charles.

Caltech, '*Which World? Scenarios for the 21st Century*, www.hf.caltech.edu-whichworld-explore-china.

China Academy of Sciences (1992), *Survival and Development – A Study of China's Long-Term Development*, The National Conditions Investigation Group Under the Chinese Academy of Sciences, Beijing, Science Press.

Cao Fenzhong (1997), 'Air & Water Quality Pollution Problems in TVEs', a policy paper prepared for NEPA, 1997.

Chen, D.W. (1994a), 'The Emergence of an Environmental Consciousness in Taiwan', Chapter 9 in Rubinstein, M.A. (ed.) (1994), *The Other Taiwan: 1945 to the Present*, Armonk, New York, M.E.Sharpe.

Chesneaux, J., Le Barbier, F. and Bergere M.-C. (1977), *China From the 1911 Revolution to Liberation*, Hassocks, Sussex, The Harvester Press (English translation of French original).

China Environmental Yearbook 1996, Beijing, China Environmental Yearbook Press (Chinese language edition).

China Statistical Information and Consultancy Centre (1999), *Statistical Communique of the People's Republic of China on the 1998 National Economic and Social Development*, Beijing: Foreign Languages Press.

CNN Earth Matters (1999), 'China's Three Gorges Dam – Eco-Boon or Cesspool?'.

Cook, I.G. and Murray G. (2000), *China's Third Revolution: Tensions in the Transition to Post-communism*, London, Curzon.

Cook, I.G. and Li, R. (1996), 'The Rise of Regionalism and the Future of China', Chapter 9 in Cook, I.G., Doel, M., and Li, R. (eds) (1996), *Fragmented Asia: Regional Integration and National Disintegration in Pacific Asia*, Aldershot, Avebury.

Cook, I.G. and Wang, Y. (1998), 'Foreign Direct Investment in China: Patterns, Processes, Prospects', Chapter 6 in Cook, I.G., Doel, M. A., Li, R. and Wang, Y. (eds) (1998), *Dynamic Asia: Business, Trade and Economic Development in Pacific Asia*, Aldershot, Ashgate.

Copper, J.F. (1996), *Taiwan: Nation-State or Province?*, Boulder, Colorado, Westview.

Dai Qing and Vermeer, E.B.(1999), 'Do Good Work, But Do not Offend the "Old Communists" ', in Draguhn, W. and Ash, R. (eds) (1999), *China's Economic Security*, London, Routledge.

Dai Qing (1998), *The River Dragon Has Come! The Three Gorges Dam and the Fate of China's Yangtze River and its People*, (edited by Thibodeau, J.G. and Williams, P.B.; translated by Yi Ming), New York, M.E.Sharpe.

Dong Liming (1985), 'Beijing: The Development of a Socialist Capital', in Sit, V.F.S. (ed.) (1985), *Chinese Cities: The Growth of the Metropolis Since 1949*, Hong Kong, Oxford University Press.

Douglas, I., Gu Hengyue and He Min (1994), 'Water Resources and Environmental Problems of China's Great Rivers', Chapter 10 in Dwyer, D. (ed.) (1994), *China: The Next Decades*, Harlow, Essex, Longman Scientific and Technical, pp. 186–202.

Dwyer, D.J. (1994), 'The Hong Kong Airport Controversy', *American Asian Review*, Vol. 12, No. 2, pp. 89–106.

Economy, Elizabeth (1998), *Case Study of China: A Summary*, Washington DC, Council on Foreign Relations

Edmonds, R.L. (1994), *Patterns of China's Lost Harmony: A Survey of the Country's Environmental Degradation and Protection*, London, Routledge.

Edmonds, R.L. (ed.) (1998), *Managing the Chinese Environment*, Oxford, Oxford University Press.

Elvin, M. and Skinner, C.W. (1974), *The Chinese City Between Two Worlds*, California, Stanford University Press.

Evans,a R. (1993), *Deng Xiaoping and the Making of Modern China*, London, Penguin Books.

Fitzgerald, C.P. (1986), *China: A Short Cultural History*, London, The Cresset Library.

Freeberne, M. (1971), 'The People's Republic of China', Chapter 5 in East, W.G., Spate, O.H.K. and Fisher, C.A. (eds) (1971), *The Changing Map of Asia: A Political Geography*, fifth edition, London, Methuen, pp. 341–447.

Freer, R. (2001), 'The Three Gorges Project on the Yangtze River in China', *Civil Engineering*, Vol. 144, February, pp. 20–8.

Gallup (1998), *Survey of Chinese Consumer Attitudes and Lifestyles*, Beijing, The Gallup Organisation.

Goudie, A. and Viles, H. (1997), *The Earth Transformed: An Introduction to Human Impacts on the Environment*, Oxford, Blackwell.

Howe, C. (1978), *The Chinese Economy: A Basic Guide*, London, Paul Elek.

Hsü, I.C.Y. (1990), *The Rise of Modern China*, New York and Oxford, fourth edition, Oxford University Press.

Hu Anyang and Wang Yi (1992), *Survival and Development – A Study of China's Long-Term Development*, The National Conditions Investigation Group Under the Chinese Academy of Sciences, Beijing, Science Press.

Huang, Jikun (2001), 'Farm Pesticides, Rice Production and the Environment', published by the Economy and Environment Program for Southeast Asia (www.eepsea.org/).

Hung, Wing-tat (1994), 'The Environment', Chapter 14 in McMillen, D.H. and Man, Si-Wai (eds) (1994), *The Other Hong Kong Report*, Hong Kong, The Chinese University Press.

Information Office of the State Council (1996), *Environmental Protection in China*, Beijing.

International Labour Organisation (1996), *Economically Active Population, 1950–2010: Vol. 1, Asia*, Geneva, ILO.

Jetsun Pema (1997), *Tibet: My Story. An Autobiography*, Shaftesbury, Dorset, Element.

Kam Wing Chan (1994), *Cities with Invisible Walls: Reinterpreting Urbanisation in Post-1949 China*, Oxford, Oxford University Press.

Kolb, A. (1971), *East Asia: Geography of a Cultural Region* (tr. C.A.M. Sym), London, Methuen.

Lao Zi (1993), *A Taoist Classic: The Book of Lao Zi* (based on Ren Jiyu's 1985 edition, translated by He, G. *et al.*), Beijing, Foreign Languages Press.

Leeming, F. (1985), *Rural China Today*, London, Longman.

Li, Choh-Ming (1967), 'Economic Development', in Schurmann, F. and Schell, O. (eds) (1967b), *China Readings 3: Communist China: Revolutionary Reconstruction and International Confrontation 1949 to the Present*, Harmondsworth, Middlesex, Penguin.

Li *et al.* (1967). *Energy Demand in China: Overview Report, Issues and Options in Greenhouse Gas Emissions Control Sub-Report Number 2*, Washington, DC, World Bank.

Li, R. (1999), 'The China Challenge: Theoretical Perspectives and Policy Implications', *Journal of Contemporary China*, Vol. 8, No. 22.

Liu Tanqi (1994), in Edmonds, R.L. (1994), *Patterns of China's Lost Harmony*, London, Routledge.

Mao Yu-Shi (1996), 'The Economic Cost of Environmental Degradation in China, A Summary', Occasional Paper for the Project on Environmental Scarcities, State Capacity and Civil Violence, Washington, DC, World Resources Institute.

Marmé, M. (1993) 'Heaven on Earth: The Rise of Suzhou, 1127–1150', Chapter 1 in Johnson, L.C. (ed.), *Cities of Jiangnan in Late Imperial China*, Albany, State University of New York Press.

Meade, R.B. (1997), 'Reservoirs and Earthquakes', Chapter 3 in Goudie, A. (ed.) (1997), *The Human Impact Reader: Readings and Case Studies*, Oxford, Blackwell.

Migot, A (1955), *Tibetan Marches* (English translation by Peter Fleming), London, Rupert Hart-Davis.

Ministry of Public Health (1996), *Selected Edition on Health Statistics of China 1991–1995*, Beijing, State Publishing House.

Muldavin, J.S.S. (1996), 'Impact of Reform on Environmental Sustainability in Rural China', *Journal of Contemporary Asia*, Vol. 26, No. 3, pp. 289–321.

——(1997), 'Environmental Degradation in Heilongjiang: Policy Reform and Agrarian Dynamics in China's New Hybrid Economy', *Annals of the Association of American Geographers*, Vol. 87, No. 4, pp. 579–613.

Murphey, R. (1980), *The Fading of the Maoist Vision: City and Country in China's Development*, London, Methuen.

Murray, G. (1993), *The Rampant Dragon*, London, Minerva Press.

——(1996), *China: The Last Great Market*, London, China Library.

——(1998), *China The Next Superpower. Dilemmas in Change and Continuity*, London, China Library.

Myrdal, J. (1967), *Report From a Chinese Village*, Harmondsworth, Middlesex, Penguin.

NEPA (1996), *National Environmental Quality Report 1991–5*.

——(1996), *Report on the State of the Environment* (Chinese language edition).

Ninth Five-Year Plan [1996–2000] for Environmental Protection and Long-Term Targets to 2010, Beijing, State Publishing House.

Nolan, P. (1990), 'China's New Development Path: Towards Capital Markets, Market Socialism or Bureaucratic Muddle?', in Nolan and Dong (eds.) (1990), *The Chinese Economy and Its Future*, Cambridge, Polity Press.

O'Riordan, T. (1981), *Environmentalism*, London, Psion.

Ögütçü, M. (1999), 'China's Energy Future and Global Implications', Chapter 5 in Draguhn, W. and Ash, R. (eds) (1999), *China's Economic Security*, London, Curzon.

Ovchinnikov, V. (1995), *The Road to Shambala*, Beijing, New World Press.

Pan, Lynn (1985), *China's Sorrow: Journeys Around the Yellow River*, London, Century Publishing.

Porter, R. (1997), *The Greatest Benefit to Mankind: A Medical History of Humanity From Antiquity to the Present*, London, HarperCollins.

Report of the Fourth National Conference on Environmental Protection 1996, Beijing, China Environmental Sciences Press.

Schafer, E.H. (1967), *Ancient China*, New York, Time–Life International.

Schurmann, F. and Schell, O. (eds) (1967a), *China Readings 1: Imperial China: The Eighteenth and Nineteenth Centuries*, Harmondsworth, Middlesex, Penguin.

Schurmann, F. and Schell, O. (eds) (1967b), *China Readings 3: Communist China: Revolutionary Reconstruction and International Confrontation 1949 to the Present*, Harmondsworth, Middlesex: Penguin.

Schurmann, H. (1966), *Ideology and Organisation in Communist China*, Berkeley and Los Angeles, University of California Press.

Shen, T.S. (2001), 'China Cannot Dismiss Pollution Hazards Haphazardly, Parts 1 & 2', *China Online*, www.ChinaOnline.com, 18/19 January 2001.

Shen Yahong (1992), 'Enlightenment From the Development of Ancient City Planning in China in the Ancient Times', *China City Planning Review*, Vol. 8, No. 1.

Simmons, I.G. (1996), *Changing the Face of the Earth: Culture, Environment, History*, Oxford, Blackwell.

Smil, V. (1993), *China's Environmental Crisis: An Inquiry Into the Limits of National Development*, Armonk, New York and London, M.E.Sharpe.

——(1996a) 'China Shoulders the Cost of Environmental Change', *Environment*, Vol. 39, No. 6.

——(1996b). *Environmental Problems in China: Estimates of Economic Costs*, East–West Center Special Reports, No. 5 (April).

Snow, E. (1970), *Red China Today: The Other Side of the River*, Harmondsworth, Middlesex, Penguin.

Spence, J.D. (1990), *The Search for Modern China*, New York and London, W.W. Norton

State Council (1996), *Environmental Protection in China*, Beijing, State Publishing House..

State Statistical Bureau, *China Statistical Yearbook 1995 and 1998*, Beijing, China Statistical Publishing House.

Takle, E. (1998), *Fuelling and Feeding an Advanced China: A Global Environmental Challenge*, Iowa State University, Iowa State University Press.

Tibet Government in Exile, *Environment and Development Issues*, Environment & Development Desk, Department of Information & International Relations, Central Tibetan Administration, Dharamsala, India, April 2001.

Thomas, D.S.G. and Middleton, N.J. (1997), 'Salinization: New Perspectives on a Major Desertification Issue', Chapter 7 in Goudie, A. (ed.) (1997), *The Human Impact Reader: Readings and Case Studies*, Oxford, Blackwell.

Tiley Chodag (1989), *Tibet: The Land and the People* (translated by W. Tailing), Beijing, New World Press.

Tregear, T.R. (1980), *China: A Geographical Survey*, London, Hodder and Stoughton.

Tuan, Y.-F. (1970), *The World's Landscapes: 1 China*, London, Longman.

United Nations (1997), *Population Division, Urban and Rural Areas 1950–2030* (1996 revision), New York, UN.

Walker, H.J. (1997), 'Man's Impact on Shorelines and Nearshore Environments: A Geomorphological Perspective', Chapter 1 in Goudie, A. (ed.) (1997), *The Human Impact Reader: Readings and Case Studies*, Oxford, Blackwell.

Wang Genxu and Cheng Guodong (2000), 'Eco-environmental changes and causative analysis in the source regions of the Yangtze and Yellow Rivers, China', *The Environmentalist*, Vol. 20, pp. 221–32.

Wang, James C.F. (1992), *Contemporary Chinese Politics: An Introduction*, New Jersey, Prentice-Hall Inc.

Wang, K. (1998), *The Classic of the Dao: A New Investigation*, Beijing, Foreign Languages Press.

Wertheim, B. (1975), *Introduction to Chinese History: From Ancient Times to the Revolution of 1912*, Boulder, Colorado, Westview Press.

Wheatley, P. (1971), *The Pivot of the Four Quarters: A Preliminary Enquiry into the Origins and Character of the Ancient Chinese City*, Edinburgh, Edinburgh University Press.

Williams, J.F. (in collaboration with Ch'ang-yi Chang) (1994), 'Paying the Price of Economic Development in Taiwan: Environmental Degradation', Chapter 8 in Rubinstein, M.A. (ed.) (1994), *The Other Taiwan: 1945 to the Present*, Armonk, New York: M.E.Sharpe.

World Bank (1999), *World Development Indicators 1997*, Washington, DC, The World Bank.

——(1997) 'Clear Water, Blue Skies: China's Environment in the New Century, Washington, DC, The World Bank.

——(1999) 'Chongqing Urban Environment Project', World Bank Project IDCN-PE-49436, Washington, DC, The World Bank.

World Resources Institute (1999). 'The Environment and China', Washington, DC, World Resources Institute.

——(1999b). 'Urban Air Pollution Risks to Children. A Global Environmental Health Indicator', Washington, DC, World Resources Institute.

——(1999c). 'China's Health and Environment: Water Scarcity, Water Pollution, and Health', Washington, DC, World Resources Institute.

Xie Qingtao, Guo Xinan and Ludwig, H.F. (1999), 'The Wanjiazhai Water Transfer Project, China: An Environmentally Integrated Water Transfer System', *The Environmentalist*, Vol. 19.

Xu Fang *et al.* (1992). 'Economic Analysis and Countermeasures: Study of TVEs Pollution's Damage to Human Health', *Journal of Hygiene Research*, Vol. 21, Supplement.

Xue Muqiao (1981), *China's Socialist Economy*, Beijing, Foreign Languages Press.

Yang, Dali (1994), 'Reform and the Restructuring of Central–Local Relations', in

Goodman, D. and Segal, G. (eds) (1994), *China Deconstructs: Politics, Trade and Regionalism*, London, Routledge.

Yabuki, S. (1995), 'China's New Political Economy: The Giant Awakes' (translated by S.M. Harner), Boulder, Colorado, Westview.

Zhang Jianguang (1994), 'Environmental Hazards in the Chinese Public's Eyes', *Risk Analysis*, Vol. 14, No. 2.

Zhang Tianlu (1997), *Population Development in Tibet and Related Issues*, Beijing, Foreign Languages Press.

Zhang, Weiping *et al.* (eds) (1994), *Twenty Years of China's Environmental Protection Administrative Management*, Beijing, China Environmental Science Press.

Zhao, Bin (1997), 'Consumerism, Confucianism, Communism. Making Sense of China Today', *New Left Review*, March–April.

Zhao, Songqiao (1994), *Geography of China: Environment, Resources, Population and Development*, New York, Wiley.

Zhou Dadi *et al.* (2000), *Developing Countries and Global Climate Change: Electric Power Options in China*, Washington, DC, Pew Centre on Global Climate Change.

Zhou Shunwu (1992), *China Provincial Geography*, Beijing, Foreign Languages Press.

Professional journals

American Asian Review
Annals of the Association of American Geographers
Atmospheric Environment
China City Planning Review
Civil Engineering
Environment
East–West Center Special Reports
International Environmental Reporter
Journal of Contemporary Asia
Journal of Contemporary China
Journal of Hygiene Research
Risk Analysis
Science
The Environmentalist
Water, Air and Soil Pollution

General media sources

Agence France Presse
Associated Press
Beijing Morning Post
Beijing Review
Beijing Weekend
Business Beijing
Business Weekly
Chinaafrica
China Central Television
China Chemical Industry News
China Daily

China Economic Times
China Environment Daily
China Environment News
China Population Today
China's Tibet
China Youth Daily
Deutsche Presse Argentur
Economic Information Daily
Guardian
Ming Pao
New York Times
People's Daily
Popular Science
Press Trust of India
Reuters
Shanghai Economic News
Shanghai Star
South China Morning Post
Straits Times, Singapore
Tibetan Bulletin
Wen Hui Bao
Xinhua News Agency

Internet sites

www.cgiar.org/cifor.html
www.China.org.cn
www.hf.caltech.edu-whichworld-explore-china
www.ChinaOnline.com
www.ciesin.org
www.eepsea.org
www.gio.gov.tw/info

Index